International Series in Operations Research & Management Science

Volume 286

More information about this series at http://www.springer.com/series/6161

Katta G. Murty
Editor

Models for Optimum Decision Making

Crude Oil Production and Refining

Springer

Editor
Katta G. Murty
Industrial and Operations Engineering
University of Michigan
Ann Arbor, MI, USA

ISSN 0884-8289 ISSN 2214-7934 (electronic)
International Series in Operations Research & Management Science
ISBN 978-3-030-40211-2 ISBN 978-3-030-40212-9 (eBook)
https://doi.org/10.1007/978-3-030-40212-9

This Springer imprint is published by the registered company Springer Nature Switzerland AG
The registered company address is: Gewerbestrasse 11, 6330 Cham, Switzerland

Preface

When I first learnt about the role of fossil fuels (coal, crude oil, natural gas) and the products made from them in our lives, I composed the following verse in Telugu:

ఫాసిల్ ఫ్లూయల్స్' అనబడే కోల్, క్రూడ్ ఆయిల్, నేచురల్ గాసులే గదా
మన ప్రస్తుత జీవన విధానానికి ముఖ్య ఆధారాలు;

వీటినుంచి మనం తయారుచేశే పెట్రోల్, డీసెల్ మన రోడ్డు వాహనాలనూ;
ఏవియేషన్ ఫ్లూయల్ ఆకాశంలో ప్రయాణం చేశే విమానాలనూ;
బంకర్ ఫ్లూయల్ సముద్రాలు, నదుల పైన ప్రయాణం చేశే ఓడలనూ, పడవలనూ
నడిపిస్తాయి;

వీటినుంచి మనం తయారుచేశే రకరకాల కెమికల్స్ ని మన దినదిన జీవనానికి
వాడుకుంటాము;

వీటినుంచి వచ్చే గాసులను మనం ఇళ్ళనూ, ఆఫీసులనూ, వేడిచేసుకోడానికీ,
చల్లబరచడానికీ,
వంటలుచేసుకోడానికీ వాడుకుంటాము;

వీటిని కాల్చి తయారుచేసే ఎలక్ట్రిసిటీ మన ఇళ్ళు, ఆఫీసులు, వీధులకు లైటు ఇస్తుంది,
మన కంప్యూటర్లు, టివిలు, పనిముట్ల నన్నిటినీ నడిపిస్తుంది;

ఇంత ముఖ్యమైన ఈ ఫాసిల్ ఫ్యూయల్స్ భూమిలో ఏర్పడడానికి మూలకారణాలైన అనాది
వృక్ష సంపదకూ, అనాది జంతు సమూహాలకూ మనం ఎంతో రుణపడిఉన్నాము.

The meaning of this verse is:

Coal, crude oil, natural gas which together are called "fossil fuels," are the corner-stones of our present way of life;

Petrol—diesel made from them are the fuels for our road vehicles; aviation fuel made from them is the fuel for our jet planes; bunker fuel obtained from them is the fuel for our ocean going ships;

We depend on many chemicals made from them in our daily living;

The gases obtained from them are used for heating or cooling our homes and offices as needed, and for cooking;

Electricity made from burning them provides lighting in our homes, offices, streets; and runs our computers, TVs, and all our implements;

We owe a huge debt of gratitude to the trees and animal groups in ancient times who led to the formation of these fossil fuel resources in the ground.

In 2015, Springer published the book "Case Studies in Operations Research: Applications of Optimum Decision Making" that I edited. Details about this book can be seen on its Springer website at:

http://www.springer.com/business+%26+management/operations+research/book/978-1-4939-1006-9.

Some readers have pointed out that the crude oil industry is the largest industry in the world today, but I only have two case studies from this industry in that book. Seeing these comments, I decided to edit this textbook as a companion volume to that book. I started working on this book during a visit to KFUPM (King Fahd University of Petroleum and Minerals) in Dhahran, Saudi Arabia between November 2016 and April 2017 on a Fulbright Scholarship.

Until then, I did not realize that the crude oil industry is the largest industry in the world today, and that many of the chemicals that we use in our daily living are made from crude oil. With the recent development of electric vehicles (road vehicles that run on electricity), and techniques to generate electricity from sunlight using solar photovoltaic cells, and windmills; some people believe that within a few years, these developments will make crude oil unnecessary. But this is not true, as we still need crude oil for all the chemicals (solvents, fertilizers, pesticides, plastics, pharmaceuticals, etc.) that we make from it to maintain our present lifestyle.

Gasoline prices in a country capture the attention of the top politicians that run the country's government. These prices are determined by the demand for gasoline and the crude oil made available in the market, by the oil-producing countries in the world. For these oil-producing countries, determining how many barrels of crude oil they should produce annually for export is an important decision-making problem with huge monetary consequences. These countries usually face pressures from superpowers to produce more crude oil to keep gasoline prices in their countries low. In this book, we consider this problem, and several other important decision-making problems that decision-makers in the crude oil industry face, and discuss procedures for finding optimum solutions for them.

Chapter 1 discusses the human use of crude oil beginning over 4000 years ago, and how over this time they developed techniques for refining crude oil into its components and make from them a variety of useful chemical products that we need in our daily living.

Chapter 2 gives brief descriptions of various operations inside a modern crude oil refinery.

In early 2015, I learnt from regular news channels that after several discussions with US Secretary of State, the authorities in Saudi Arabia have decided to increase their crude oil exports from 9.6 million barrels/day to 10.6 million barrels/day. I was very surprised to see that within a matter of days, the market price of crude oil fell from $100–110/barrel to $50 /barrel, and kept falling still further; resulting in a huge decrease in the revenues of crude oil-exporting countries from the sale of their crude oil. Then I realized that the problem "How much crude oil should a crude oil-exporting country produce for export annually" is the decision-making problem with the largest monetary consequences. Chapter 3 presents possible solutions to this problem. We discuss the important objective functions they need to consider in making this critical decision, and discuss procedures to find the best solution.

During my research in earlier years, I already became acquainted with offshore drilling for crude oil and learnt that the operations involved in moving drilling rigs from one location to another, and horizontal drilling over long distances, are both very expensive. The important decision-making problem in offshore drilling is to allocate the various target locations in the field, the available drilling rigs to drill. In essence, this decision requires partitioning the set of target locations to drill, into subsets or clusters, each subset to be drilled by one of the available rigs; to minimize the total cost of drilling all the target locations. In the literature, a decision-making problem like this is known as a "clustering problem." Finally, I was able to develop an approach for this problem based on the Shortest Hamiltonian Path problem, and a friend at the University of Waterloo told us that they have the efficient Concorde TSP code to solve it. And a Petroleum Engineer at Aramco provided realistic data for an application in Saudi Arabia. We worked together to solve this problem and prepared Chapter 4 based on it. We believe that this approach gives better results for this problem, than other approaches for it in the literature.

Chapter 5 discusses a very important decision for crude oil producing and exporting countries like Saudi Arabia. They can either export the crude oil that they produce directly; or set up oil refineries in their country for refining the crude oil they produce and sell the products that this produces. The second alternative not only leads to much higher profits but also has the additional advantage of providing jobs for several of their countrymen. Based on simple product mix examples in oil refining, Saudi Arabia realized this in early 2000s and has since been focussing on investing in refining much of the crude that they produce. We discuss the optimum product mix problem for an oil refinery in this chapter.

Outputs from the treatment units in an oil refinery are only semifinished products. They are blended into finished products like gasoline, diesel oil, etc., meeting various specifications that the marketplace demands. Chapter 6 discusses this product blending problem.

Chapter 7 discusses the problem of blending available crude oils to produce an optimal mix that is the feedstock into the crude distillation unit, in the refinery.

Chapter 8 discusses the problem of allocating pumps to blend component materials from component tanks into a final product tank to ensure that this final blending step is carried out as quickly as possible.

We (my co-authors and I) hope that before these wonderful natural resources are gone, they will be used for the maximum benefit to humankind.

Ann Arbor, USA Katta G. Murty

Acknowledgements

My thanks to all the following people and organizations:

The Fulbright Organization for their scholarship to visit KFUPM (King Fahd University of Petroleum and Minerals) in Dhahran, Saudi Arabia, which got me started in preparing this book;

Contributing authors Ahmed Alsaati, Geoffrey Gill, Shane G. Henderson, David Kaufman, and Parvati Patel for their collaboration;

William J. Cook of University of Waterloo for giving me information on their downloadable Concorde TSP software for solving the Hamiltonian Path problem in Chap. 4;

Gregory Shahnovsky, Ariel Kigel, and Ronny McMurray of Modcon Systems, UK, for their paper [1] with information and realistic data for preparing Chap. 7;

Geoffrey Gill for his insightful feedback on the first seven chapters;

My friends Bala Guthy, Satya Kodali, my daughter Madhusri Katta, and her Husband Jason Kerr, who have read preliminary drafts and gave me comments to improve them;

Sasanka Mouli Neti for solving the LP model in Chap. 5;

Edwin Tang from my department for helping me with downloading the figures; and

The ISOR My Springer Editor, Camille Price, and the Springer team for constant encouragement.

The figures in this book are taken from public web sources from Wikipedia articles, https://commons.wikimedia.org, and https://pixabay.com/images/search/oil%20refinery/, and we gratefully acknowledge the owners.

Finally, I thank my Wife Vijaya Katta for being a great companion.

October 2019

Katta G. Murty
murty@umich.edu

Reference

1. "The Economics of Process Analysis", Gregory Shahnovsky, Ariel Kigel, and Ronny Mc-Murray of Modcon Systems, UK; Hydrocarbon Engineering, (85–92), February 2017, paper can be accessed at the link: https://www.modcon-systems.com/wp-content/uploads/2014/04/The-Economics-of-Process-Analysis.pdf.

Contents

Contributors

Ahmed AlSaati Well Completion Operations and Production Engineering Department, Aramco, Dhahran, Saudi Arabia

Geoffrey Gill Viva Energy Refining Pty Ltd, Corio, VIC, Australia

Shane G. Henderson School of Operations Research and Information Engineering, Cornell University, Ithaca, NY, USA

David Kaufman Decision Sciences, College of Business, University of Michigan, Dearborn, MI, USA

Katta G. Murty Department of IOE, University of Michigan, Ann Arbor, MI, USA

Parvati Patel Hamilton City Council, Hamilton, New Zealand

Chapter 1
History of Crude Oil Refining

Katta G. Murty

Abstract In this chapter, we discuss the history of the use of crude oil by humans in various areas of the world from more than 4000 years ago, and how crude oil refining has evolved over time, and the various products that we make from crude oil for use in our modern lifestyles.

"Crude oil" or "petroleum" (from Greek: "petra" ("rock") and "oleum" ("oil")) is a naturally occurring yellow to black liquid found in geological rock formations beneath the earth's surface. It is a "fossil fuel" formed in very ancient times when large quantities of dead organisms were buried underneath sedimentary rock and subjected to both intense pressure and heat underground. For a very long period, people in many different parts of the world have been noticing natural seepage of subsurface crude oil collecting in oil pits. In [1, 2] it was reported that oil collected by hand from such pits was used more than 4000 years ago in Babylon in the construction of walls and towers, asphalt used to seal water tanks at Mohenjo-Daro [3], in Ancient Persia for medicinal and lighting uses, and also in some parts of the Roman Empire. The earliest known oil wells were reported to have been drilled using bamboo poles in China by the fourth century CE. By the seventh century CE, the Japanese statesman **Shen Kuo** of the Song Dynasty coined the word **rock oil** for crude oil.

Crude oil is flammable, so people tried to use it for lighting. But the burning crude produced an awful smell and a great deal of smoke, so it was hard to sell as an illuminant. So, they realized that it needs to be refined by distillation to get the best lamp oil from it. Crude oil is a mixture of a vast number of hydrocarbon compounds ranging from the smallest fractions such as methane or natural gas to large asphaltenes and paraffin chains; these can be separated because they have different boiling points. The lightest molecules have the lowest boiling points and the heaviest ones the highest. Generally the more carbon atoms in the molecules of a crude oil compound, the higher its boiling temperature and density. Examples: Propane

K. G. Murty (✉)
Department of IOE, University of Michigan, Ann Arbor, MI 48109-2117, USA
e-mail: murty@umich.edu

K. G. Murty (ed.), *Models for Optimum Decision Making*, International Series in Operations Research & Management Science 286, https://doi.org/10.1007/978-3-030-40212-9_1

(C_3H_8), Butane (C_4, H_{10}), Decane ($C_{10}H_{22}$) have boiling temperatures −43.6 °F, 30.2 °F, 345 °F and densities 1.74, 2.48, 730 kg/m^3, respectively.

Every oil field produces crude oil with a different mixture of hydrocarbons. Crude oil is classified by density into three categories **light, medium, and heavy**. It is "light" ("heavy") if it has a higher proportion of "lighter" ("heavier") compounds. Compounds used in gasoline blends are lighter hydrocarbons (these are hydrocarbons with molecular weight less than that of heptane (C_7H_{16})), whereas those used for making lubricating oils, wax, and asphalt are heavier hydrocarbons.

Crude oil is classified as **sweet** if it is low in sulfur (less than 1% sulfur content by weight), **sour** if it is high in sulfur (may contain 1–5% sulfur).

The terms **low (heavy) ends** refer to "lower (higher)" boiling point components in a mixture of hydrocarbons. At 90 °F, the lightest fractions of crude oil vaporize; these are gases such as methane, ethane, and butane. Naphtha, which is a compound in the gasoline blend, is vaporized between 90 and 220 °F, Kerosene is vaporized between 315 and 450 °F, gasoil (this generally gets refined into diesel) between 450 and 650 °F, and fuel oil between 450 and 800 °F. **Light oils** refers to products distilled from crude oil up to but not including the first lubricating oil distillate. **Heavy fuel** oil is fuel oil having a high density and viscosity. When the temperature of crude oil raises above 1000 °F, heavy hydrocarbons in it begin to "crack" (i.e., break up into lighter compounds). This **cracking** feature was not part of "straight-run distillation" used in earlier times, it was unknown until modern times.

Crude oil accessible from natural fields was being distilled in oil refineries for the production of chemicals like tar, and flammable substances like kerosene in Persia, Arab countries, Europe, and Russia by the thirteenth century **CE (Current Era)**.

These first oil refineries used **straight-run distillation** (the first form of **atmospheric distillation** used), the simplest distillation method, to extract kerosene, the most valuable product until 1915; and other light fractions like naphtha and gasoline. Compared to oil refineries today, these early refineries were tiny; **they distilled crude oil in batches** in horizontal cylindrical iron stills that only hold 5–6 barrels at a time, built over a furnace. They would raise the temperature of the oil very slowly. The vapors would be captured in a condenser and allowed to condense back into liquid. After each fraction of crude oil is boiled off, the still was heated further until the next fraction boils. The operator would control the product coming out of the condenser by controlling the temperature of the still. Each fraction would be collected in a separate container, and possibly redistilled to produce a purer fraction (see [4, 5]).

The left over residue, approximately half the volume of the input crude oil, was too heavy to vaporize in the process used; it was stored in tanks until it was shipped out to be processed into lubricating oils and greases, bitumen, or sold as fuel oil. There was no market for gasoline or naphtha at that time, and their explosiveness made their storage a problem, so they were simply poured into open pits and burned, or disposed off in a nearby river. The very lightest fractions like methane, butane, and propane were either flared off or escaped into the atmosphere.

By the thirteenth century, crude oil was being distilled in this way in the Arab world, Persia, Europe, Russia, China, and other parts of Asia to make flammable products and other chemicals like lubricants. Early British explorers to Myanmar documented a flourishing oil extraction industry based in Yenangyaung that in 1795 CE had many hand-dug wells under production.

In those days, refiners used to sell their product to "jobbers" who would run retail operations. The jobbers would deliver kerosene from a tank on the back of a wagon to homes and businesses. Hardware and grocery stores stored barrels of kerosene and other petroleum products in the back of the store; customers would bring their own containers to be filled from the barrel.

In the mid-eighteenth century under the Empress Elisabeth of Russia, a refinery was built in Russia. Through this process of distillation of the rock oil they produced a kerosene-like substance which was used in oil lamps in Russian churches and monasteries, even though households still relied on candles. By 1825, Russia was producing and processing 3500 barrels/day of oil this way.

Modern oil drilling began in Azerbaijan in 1848, and two large **pipelines** were built in the Russian empire: the 833 km long Baku–Batumi pipeline to transport oil from the Caspian to the Black Sea Port of Batumi completed in 1906, and the 162 km long pipeline to carry oil from Chechnya to the Caspian. At the turn of the twentieth century, Russia's output of crude oil accounted for half of the world's production and dominated international markets.

The first modern oil refinery based on **continuous distillation** was built by Ignacy Lukasiewicz near Jasha in Poland in 1854–56. Compared to refineries of today, it was a small refinery, but built according to modern designs based on **fractional distillation**. The refined products used were kerosene, asphalt, fuel oil, and lubricants. As Lukasiewicz's Kerosene lamps gained popularity, the refining industry grew in the area.

The name **kerosene** (derived from the Greek word "keroselaion" meaning "wax-oil") for the product was coined by the Canadian geologist **Abraham Pineo Gesner**; he developed the process of making it from coal, bitumen, and oil shale. In 1850 he created the **Kerosene Gas Light Company**, and began installing lighting in the streets of Halifax and other cities in Canada.

The first commercial oil well in North America was drilled in Canada by **James Miller Williams**, and it started producing oil in 1858. He discovered a rich reserve of oil four meters below ground. In 1862 Canada's oil man **John Shaw** stuck oil at 48 m leading to the first major oil gusher coming into production, shooting crude oil into the air at a recorded rate 3000 barrels/day [1].

In 1859, two separate events occurred that would jumpstart both the **petroleum** and the **auto industries**. In that year, **Edwin L. Drake** in the US drilled the **first working oil well in the US** in Titusville, Pennsylvania (a 70-foot deep well) at a cost of $15000 (Nelson, 1958 [6]); and the French engineer **J. J. Etienne Lenoir** made the first dependable internal combustion engine powered by gasoline. Drake's oil well kicked off the petroleum industry, and Lenoire's work lead to the development of the automobile of today (see [7]). Between 1860 and 1900 CE, many technological innovations occurred in the developing oil and auto industries. By 1908, the year that **Henry Ford's Model T** made its market debut, the US had roughly 125000 passenger cars. By 1911 gasoline dethroned kerosene as the top selling product for Standard Oil of New Jersey, the largest refiner in the US at that time. Kerosene's slide was hastened by the 1910 invention of the tungsten filament for electric light bulbs by **William David Coolidge** (the tungsten filament outlasted all other types of filaments, and Coolidge made the costs practical). Drake's first refinery, and others around that time, used batch distillation to separate kerosene and heating oil from other crude fractions. However with increasing demand for petroleum products, continuous refining became a necessity, and the first **continuous refinery plants** emerged around 1912 (Nelson, 1958 [6]).

In 1878 **Ludwig Nobel** and his Branobel company revolutionized oil transport by commissioning the **first oil tanker** and launching it on the Caspian sea [1]. It was built in Sweden and operated between Baku and Astrakhan.

On January 10, 1901, a US Mechanical Engineer named **Anthony Lucas** struck crude oil beneath a dome of land called **Spindletop Hill** a few miles south of Beaumont, Texas. The site produced a gusher unlike any the world has seen before, spewing more than 100 feet into the air continuously for 9 days before Lucas and his men could bring it under control. Other prospectors began spending huge sums to acquire nearby lands and sink drills of their own.

More gushers soon followed. Before long, The **Texas oil boom** was on, and drillers came to Texas from all over. By 1910 crude oil production within the US was more than equalled to that of the rest of the world combined. So, in the first quarter of the twentieth century, the US overtook Russia as the world's largest oil producer. By the end of this quarter, oil fields had been established in many countries including Canada, Poland, Sweden, Ukraine, US, Peru, and Venezuela.

In 1913 refiners developed **thermal cracking** which enabled production from a barrel of crude oil, of more gasoline, diesel fuel, and kerosene which has become a primary component in aviation fuel for modern jet engines, and soon several other process innovations followed, allowing refiners to meet market needs.

The products refined from the liquid fractions of crude oil today can be placed into 10 major categories. These are (see [5]) **asphalt** (used for paving roads), **diesel** (produced in fractional distillation between 392 and 662 °F, has higher density than gasoline, burns at higher temperature than gasoline, evaporates more slowly, and commonly used in transportation as diesel engines are more efficient than gasoline engines), **fuel oil** (heaviest commercial fuel produced in refineries, burned in furnaces to generate heat, part of it is used as bunker fuel for ocean-going vessels), **gasoline** (almost 50% of output from refineries, it is a blend of paraffins, naphthenes, olefins, etc., mainly used as fuel for internal combustion engines), **kerosene** (collected in fractional distillation between 302 and 527 °F, used as jet fuel and heating fuel; it is thin, clear, and burns without any smell as it contains no sulfur compounds), **LPG** or **Liquified Petroleum Gas** (includes propane, butane; these are a mixture of gases used in heating appliances, refrigerants, and aerosol propellants; since these are gases at normal atmospheric pressures, they are marketed in pressurized steel bottles), **lubricating oils** (used between two surfaces to reduce friction and wear, an example is motor oil), **paraffin wax** (excellent electrical insulator, also used in drywall to insulate buildings, and in candles), **bitumen** (also known as tar, used in asphalt, is from the bottom fraction in fractional distillation, used in paving roads, waterproofing roofs, also used as thin plates to soundproof dishwashers), and other petrochemicals (includes **ethylene** used to make anesthetics and plastics like **polyethylene, antifreeze**, and **detergents, propylene, benzene, toluene, xylene** used in making many different chemical products for our daily use).

Crude oil and liquid products from oil processing are usually measured in barrels. A **barrel** is 42 gallons. In oil refining, from a 42-gallon barrel of input crude oil, the output is roughly 45 gallons of petroleum products. This is due to the overall density changes that occur in cracking operations in the refining process.

Example of Higher Volume Outputs From Cracking: When some of the straight-run heavy gas oil fractions from crude oil are heated to temperatures of 900 °F or higher, the molecules in them begin to crack into two or more smaller molecules. This phenomenon is called **cracking**. We give an example of that here.

Cetane ($C_{16}H_{34}$) is a product produced in crude oil refining. It weighs 7.2 lbs/gallon. When a unit weight of Cetane is cracked, it produces 50% Octane (C_8H_{18}), 38% Hexene (C_6H_{12}), and 12% Ethylene (C_2H_4), by weight. Octane, Hexene, and Ethylene have densities 5.9 lbs/gallon, 5.6 lbs/gallon, and 3.1 lbs/gallon, respectively. So, when 1 gallon of Cetane is cracked, it results in 0.61 gallons, 0.48 gallons, 0.29 gallons of Octane, Hexene, and Ethylene, respectively, for a total of 1.38 gallons of cracked products [8].

Refiners use cracking in controlled ways extensively. In this process, straight-run heavy gas oils are subjected to heat (900 °F or higher) and pressure in the presence of a catalyst to promote cracking. This process helps convert those heavy low-value products into lighter components in high-value products like gasoline blends.

Today crude oil refineries are operating all over the world. The **Jamnagar Refinery** of Reliance Industries in India refining 1240000 barrels/day, **SK Energy** in Ulsan, South Korea processing 1120000 barrels/day, and the **Paraguana Refinery complex** in Venezuela processing 940000 barrels/day are the world's largest. The top 11 of the world's largest refineries are located in India, South Korea, Venezuela, Singapore, Texas in the US, Saudi Arabia, and Greece.

For clean performance when they are in use, for each product stringent specifications have been developed on several important performance characteristics. As an example, we will explain the function of the specifications on the characteristic **octane rating** for gasoline (see [9–11]). Gasoline must meet three primary requirements—it must have an even combustion pattern, start easily in cold weather, and meet prevailing environmental requirements. Gasoline must burn smoothly in the automobile engine, otherwise "**knocking**", "severe knocking" can occur jolting passengers and damaging car engine. Experiments determined that the most severe knocking occurs with fuel of pure normal heptane, while least knocking was produced by pure isooctane. This led to the development of the **octane scale** for defining gasoline quality. When a motor gasoline gives same knocking performance as a mixture of 90% isooctane and 10% normal heptane, it is given the octane rating of 90. Adding oxygenated compounds such as ethyl alcohol to the gasoline blend also helps to increase its octane rating and also to reduce emissions of carbon monoxide and nitrogen oxides. Gasoline production requires a special process termed **reforming** to meet octane specifications.

Octane rating or octane number is not an indicator of the energy content of the fuel. It is only a measure of the fuel's tendency to burn in a controlled manner, rather than exploding in an uncontrolled manner. It is possible for a fuel to have **Research Octane Number** (**RON**, most commonly used octane rating worldwide) more than 100 because isooctane is not the most knock-resistant fuel available. For example, racing fuels have octane ratings of 110 or greater.

Most gas stations offer three octane grades: **regular** (usually 87 octane), **mid-grade** (usually 89 octane), and **premium** (usually 92 or 93). Regular octane is recommended for most cars. Some cars with high compression engines (like sports cars, luxury cars) need mid-grade or premium-grade to prevent knocking. It is interesting to note that in Australia today, regular, mid-grade, and premium gasoline octane ratings are 91, 95, and 98. There are three different estimates used for measuring the octane rating of gasoline; they are RON (Research Octane Number), MON (Motor Octane Number), and (RON + MON)/2. The differences in the octane ratings of the three grades (regular, mid-grade, and premium grade) in the USA and Australia may be due to the two countries using different estimates among RON, MON, (RON + MON)/2 for the octane rating specifications in their countries.

During the 1970s and 1980s, farm lobbies around the world successfully convinced their Governments that including ethanol made from fermenting corn, sugar cane, and some other crops, with octane rating of 114, in gasoline blends to enhance its octane rating, would be a good national strategy. Government subsidies brought on large volumes of this natural ethanol capacity, and from it **gasahol** (a blend of 90% conventional motor gasoline made from crude oil, and 10% ethanol) entered the market.

In the twentieth century, oil companies prospered along with automakers. At the beginning of this century, ocean-going steamships converted from using coal as their primary fuel, to using heavy fuel oil (under the name **bunker fuel**), which gave the big advantage of being pumped into the engine rather than being shoveled. Refiners built more refineries and expanded existing facilities. Researchers improved thermal cracking techniques and developed other catalytic processes to produce high-grade products. High-octane gasoline emerged in the 1930s from these efforts. These developments played a major role in WWII (World War II), in which US oil refiners supplied more than 80% of the fuel used by allied forces. Running on high-octane fuel that optimized engine performance, the equipment of allied forces in WWII outmaneuvered German equipment using inferior fuels. **Joseph Stalin** of the Soviet Union said in a toast during a banquet toward the end of the war: "This is a war of engines and octanes. I drink to the American Auto and oil industries". See [7]. Another important development was the building of **long-distance pipelines**, which had overtaken rail tank cars as the main method of transporting hydrocarbons. One example is **The Big Inch** spanning 1254 miles for moving crude from Texas oil fields to East Coast refineries built in 1942, 1943 (world's longest crude oil pipeline at that time), followed by the **Little Inch** which transported refined products.

After the end of WWII, the countries in the Middle East took the lead in crude oil production from the US. There were several important developments after WWII, including ocean and deep-water drilling, growth of the global shipping network for crude oil relying on huge oil tankers and pipelines, international organizations like **OPEC** (Organization of the Petroleum Exporting Countries) playing a major role in setting petroleum prices and policy. With consumption over 30 billion barrels/year, **crude oil refining** is one of the major industries all over the world today. Crude oil refineries process crude oil into transportation fuels (gasoline for automobiles, diesel fuel for trucks and railcars, aviation fuel for jet planes, and bunker fuel for ships) that we need for our mobility all over the planet; lubricating oils and many chemicals that we need to maintain our present lifestyle. Today crude oil has become an essential commodity for the sustainability of our modern way of life; its importance is due to all the machines based on internal combustion engines that we use in our daily living, rise of commercial aviation, and to industrial organic chemistry in manufacturing fertilizers, pesticides, plastics, solvents, adhesives, and so many other chemicals that we use daily. Today crude oil refining is the most prominent and complex industry in the world, offering many challenging applications of optimization techniques [12]. A typical refinery has many different crude oils and other feedstocks to choose from, dozens of different products to manufacture to maximize profit based on consumer demands in the global marketplace, and hundreds of important decisions to be made in daily operations. But the crude oil industry faces serious environmental concerns due to **GHG (green house gas)** emissions into the atmosphere from the burning of fossil fuels, and the consequent global warming; and oil spills and their cleanup.

The main type of bunkerfuel used by ships is heavy fuel oil which contains sulfur, which after combustion in the engine, ends up in ship emissions. These emissions containing sulfur oxides are known to be harmful to human health. In the atmosphere these sulfur oxides lead to **acid rain** which harms crops, forests and aquatic species, and contributes to the **acidification of the oceans**. So the **IMO (International Maritime Organization)** introduced regulations in 1997 to reduce sulfur oxides emissions from ships to 0.5% by weight, from 2020, this is known as **MARPOL 2020**. This poses a challenge to producers of crude oil with high sulfur content, refiners, and shipping companies.

In 2009, the total world crude oil reserves have been estimated to be 1342 billion barrels, with Venezuela as the oil kingpin in OPEC with 301 billion barrels of proved oil reserves (24.8% of OPEC's total), and Saudi Arabia as the country with the second highest reserves of 267 billion barrels (22% of OPEC's reserves). The high monetary value of crude oil and products made from it, have led to it being labeled as **black gold**. And this black gold continues to gush forth from the ground in millions upon millions of barrels, leading to the rise of the **Automobile age**, two world wars, and the metamorphosis of **USA into a Superpower**.

Vertical Depths of Crude Oil Wells: The first crude oil well drilled in Texas in 1866 was a little over 100 feet deep. Data from Energy Industry Administration shows that in 2009, the average depth of US exploratory wells was almost 7800 feet. The deepest crude oil well is BP's Tiberfield in the Gulf of Mexico, its depth is over 35,000 feet.

Optimum Decision-Making (also known as "Mathematical Programming") is a branch of the subject "Operations Research (OR)", which has a rich history in the chemical and petroleum industries. These industries were early adopters of OR, especially optimum decision-making, and in turn contributed to the development of OR as a subject.

References

1. "History of the Petroleum Industry", from Wikipedia, see https://en.wikipedia.org/wiki/History_of_the_petroleum_industry. Also see "Petroleum" at https://en.wikipedia.org/wiki/Petroleum.
2. "Petroleum Industry", from Wikipedia, see https://en.wikipedia.org/wiki/Petroleum_industry.
3. "Petroleum Refining Processes", Jamed G. Speight and Baki Ozum, Marcel Dekker, 2002.
4. Aspen Junge, "Petroleum Refining: A 125 Year Kansas Legacy", Kansas Dept. of Health & Environment, see www.kdheks.gov/remedial/articles/refining_history.pdf.
5. "Refining Crude Oil: History, Process and Products", see www.oil150.com/files/refining-crude-oil-history,-process-and-products.pdf.
6. "Petroleum Refinery Engineering", W. L. Nelson, McGraw-Hill, NY, 4th Ed., 1958.
7. "How Oil Refining Transformed US History & Way of Life", Industry Market Trends, see https://news.thomasnet.com/imt/2003/01/17/how_oil_refinin.
8. BP Statistical Review of World Energy 2018, https://www.bp.com/content/dam/bp/en/corporate/pdf/energy-economics/statistical-review/bp-stats-review-2018-full-report.pdf.
9. "Petroleum Refining", Encyclopedia Britannica, see www.britannica.com/technology/petroleum-refining.
10. "Oil_ Refinery", from Wikipedia, see https://en.wikipedia.org/wiki/Oil_refinery.
11. " Petroleum Refining in Nontechnical Language", William L. Leffler, 4th Edition, Pennwell Corporation, 2010.
12. "Case Studies in Operations Research: Applications of Optimum Decision Making", Katta G. Murty (editor), Springer, 2015.

Chapter 2
Operations Inside a Crude Oil Refinery

Katta G. Murty

Abstract In this chapter, we give brief descriptions of various operations inside a modern crude oil refinery, and the immense benefits that we derive from products made out of crude oil.

Crude oil is either locally produced or imported. Local crude comes from production sites which may be either offshore or onshore. Offshore crude oil production wells are drilled from either floating (discussed in Chap. 13 of [1]) or fixed platforms that are placed on the ocean floor. Offshore production of crude oil is transported either by pipelines or tankers to maritime terminals. From there, it is either exported or shipped to refineries for processing.

Fixed platforms on the ocean floor contain storage tanks for gas and oil, as well as pumps to transport the resource by pipeline, or to fill tanker ships. They may contain bedrooms and lounges for up to 200 workers. The power supply for the many items of equipment, and for the living quarters is produced by generators on the fixed platform.

When crude oil is extracted from a reservoir, a little over 50% of the crude oil in the reservoir still remains in the ground, increasing the world crude oil recovery rate from reservoirs by 5% would result in more new oil than all of Saudi Arabia's current fields.

Typical production fluids coming from the oil wells are a mixture of crude oil, gas, and produced water. Many permanent offshore platforms have full oil production facilities on board. Smaller platforms and subsea wells export raw production fluid to the nearest production facility which may be an onshore terminal or a nearby offshore processing platform.

The processing consists of a separator to separate the three components: crude oil, gas, and water in the mixture. The separated oil is usually routed to a coalescer

K. G. Murty (✉)
Department of IOE, University of Michigan, Ann Arbor, MI 48109-2117, USA
e-mail: murty@umich.edu
URL: http://www-personal.umich.edu/~murty/

K. G. Murty (ed.), *Models for Optimum Decision Making*, International Series in Operations Research & Management Science 286, https://doi.org/10.1007/978-3-030-40212-9_2

Fig. 2.1 Conoco oil refinery. Figure from https://commons.wikimedia.org

before being metered and pumped to an onshore terminal. The produced water goes through treatment to remove entrained oil and solids, and then either reinjected into the reservoir or dumped overboard. The produced gas (dubbed "wet gas") is treated to remove the liquids, and the resulting "dry gas" is either exported by pipeline, or reinjected into the reservoir, or used as fuel for installation's power generation, or flared.

Modern Petroleum Refineries are typically large sprawling industrial complexes with tubes of various diameters running throughout (Figs. 2.2, 2.3 and 2.5). These tubes carry different streams of fluids between various processing units in the refinery. Other distinct signs of an oil refinery are processing towers of various heights (like the Fractionation Tower for fractional distillation described in Chap. 11 of [1]) and farms of many storage tanks for storing crude oil until the refinery is ready to process it, and storing various liquid products from the refinery (Figs. 2.1, 2.2, 2.4, 2.6 and 2.7). Unless maintenance is needed, refineries operate round the clock every day.

Crude oil refineries are the main part of the downstream side of the petroleum industry; they are typically large scale plants processing hundred thousand to several hundred thousand barrels of crude oil/day. Because of the high capacity of these plants, making decisions in the daily operations in them optimally can lead to substantial savings.

Oil refineries use a lot of steam and cooling water, so they have to be located at a site with an abundant source of water, like near a navigable river, or seashore near a port. And their major outputs are transported by river, or sea, or by pipeline. Small volume outputs are transported by railcars or road tankers. So the site of an oil refinery should have infrastructure for receiving supply of raw materials, and for shipment of products to markets.

Also, since the refining process releases several chemicals into the air, and polluting wastewater on the ground, and poses risks of fire, explosions, and harmful release of chemicals; residential communities vigorously oppose the siting of a new

Fig. 2.2 In this oil refinery aerial photo from https://pixabay.com/images/search/oil%20refinery/, you can see refinery units and oil storage tanks

oil refinery close to their neighborhood. So, typically oil refineries have to be located some distance away from populated areas.

The Crude Distillation Unit (CDU) described in Chap. 11 of [1], also referred to as the *atmospheric distillation unit* since it operates at slightly above the atmospheric pressure, is the first processing unit in all oil refineries; it separates the crude oil into various fractions in it of different boiling ranges, each of which undergoes additional processing in other processing units in the refinery.

Based on the way crude oil is distilled and separated into fractions, the products from a refinery are usually grouped into these categories:

Light distillates: The gaseous and more volatile liquid hydrocarbons produced in an refinery are collectively known as light hydrocarbons or lightends. This group includes LPG, gasoline, and naphtha; these fractions have the lowest densities and boiling points.

Middle distillates: Kerosene, diesel: these are fractions with boiling points and densities between those of kerosene and lubricating oil fractions.

Heavy distillates: Heavy fuel oil, lubricants, wax, asphalt, and tar; these fractions have the highest boiling points and densities.

The middle distillates result in the products: diesel, marine gasoil, and jet fuel, and are used in road transportation, shipping, aviation, and manufacturing; they account for more than a third of global oil consumption. In 2017 middle distillates consumption globally was 35.3 million barrels per day or 36% of the world's crude oil consumption of 98.2 million barrels per day according to the BP Statistical Review of World Energy 2018.

Other output products of oil refining are hydrogen, and light hydrocarbons which are steam-cracked in an ethylene plant, and the produced ethylene is polymerized to produce polyethylene.

The refining process actually begins at the oil well itself. The pumped oil comes mixed with brine (salt water) and mud as it comes out of the well. In the settling tanks, the oil, brine and mud, settle into layers. The crude oil is pumped from the top of the tank into a pipeline or truck for delivery to a refinery. The brine and mud are injected right back into the oil-bearing geologic formation to repressurize the formation to make the oil easier to pump.

Refinery processes include: **distillation** (to separate crude oil into the various products contained in it of different boiling points), **cracking** (the products obtained from distillation that have high boiling points and densities are cracked which breaks heavy hydrocarbon molecules into lighter ones). Cracking units are classified into hydrocracking, fluid catalytic cracking, coking, **hydrotreating** (this purifies products from distillation and cracking to remove most of their sulfur and increase hydrogen content. Hydrogen is either purchased or manufactured for this), **reforming** (in gasoline production, this process improves its octane rating to its specified level).

Here are various processing units in a typical oil refinery [2, 3]. In modern oil refineries, these process units are combined into functional categories: crude supply and blending (includes receiving facilities, tank farm for receiving and storage, and either blending, or sent directly to the processing system), separation (basically distillation to separate various fractions), upgrade, treatment, and other processing and blending according to product specifications. This portion also includes utilities that provide the refinery with power, fuel, steam, cooling water, compressed air, etc., all necessary things for operations in the refinery. And, finally dispatch of finished products.

1. **Desalter Unit:** This unit is designed to wash salt out from crude oil before it enters the CDU. It mixes crude oil with water; the salt will dissolve in water, and as the water and oil are separated, the salt is washed away.
2. **Atmospheric Distillation Unit:** Also called CDU (Crude Distillation Unit), this unit separates the crude oil into its various fractions.
   Asphalt is one of the heavier hydrocarbons that settles in one of the trays near the bottom in the Fractionation Towers in the CDU, it is sent for storage as asphalt. The remaining residual bottom is sent to the Vacuum Distillation Unit described below.
3. **Vacuum Distillation Unit:** This unit (also called **Vacuum Flasher**) further distils residual bottoms from the CDU. Then **Solvent Dewaxing Units** remove the heavy waxy constituent petrolatum from Vacuum Distillation production.
   The last portion that is drawn off the Vacuum Flasher is called **Flasher Bottoms**; some portion of this may be used as asphalt, and the rest sent on to further processing.
4. **Naphtha Hydrotreater Unit:** In this unit, hydrogen is used to desulfurize naphtha from the CDU before it is sent to the catalytic reformer unit.
5. **Catalytic Reformer Unit:** This unit converts the naphtha evaporating range products (see Table 11.1 in Sect. 1.2.2 in Chap. 11 of [1]) into higher octane reformate which contains higher proportion of aromatics and cyclic hydrocarbons. This unit also produces hydrogen as a byproduct; this hydrogen is used in hydrotreaters

(described in no. 4 in this list), or the hydrocrackers (described in no. 8 in this list).

6. **Distillate Hydrotreater:** This unit desulfurizes distillates like diesel from the CDU.
7. **Fluid Catalytic Cracker (FCC) Unit:** This unit upgrades heavier fractions from the CDU, into more income producing lighter products, or asphalt.
8. **Hydrocracker Unit:** This unit uses the hydrogen from no. 5 in this list to also upgrade heavier fractions from the CDU into lighter, more profitable products.
9. **Visbreaking Unit:** This unit is used to convert heavy residual oils from the CDU by thermally cracking them into lighter, more profitable reduced viscosity and pour point products.
10. **Merox Unit:** This unit is used to treat LPG, kerosene, jet fuel, to oxidize mercaptans in them into organic disulfides. Other units (**Doctor sweetening process, Caustic washing**) are also used for this purpose.
11. **Coking Units:** There are three types of coking units, Delayed Coking, Fluid Coker, and Flexicoker. These units process heavy residual oils into gasoline and diesel fuel, leaving Petroleum Coke as a byproduct.
12. **Dimerization Unit:** This unit converts olefins into higher-octane gasoline blending components. Butenes can be dimerized into isooctene which may subsequently be hydrogenated to form isooctane.

    Alkylation Unit uses sulfuric acid or hydrofluoric acid to produce high-octane components for gasoline blending.

    Isomerization Unit converts linear molecules into higher-octane branched molecules for blending into gasoline or feed to Alkylation units.

13: **Steam Reforming Unit:** This unit produces hydrogen for the hydrotreaters, hydrocrackers (discussed in nos. 4, 8 in this list).

14: **Amine Gas Treater, Clause Unit, and Tail Gas Treatment:** These convert hydrogen sulfide from hydrodesulfurization in no. 6 in this list, into elemental sulfur.

15. **Solvent Reforming Units:** These units use solvent (cresol or furfural) to remove unwanted aromatics from lubricating oil stock or diesel stock.

    Solvent Dewaxing Units remove the heavy waxy constituents petrolatum from vacuum distillation products from no. 3 in this list.

A Schematic flow diagram of a typical oil refinery is given in Fig. 2.8.

The refinery also has **Utility units**; these are **cooling towers** circulating cooling water, **Boiler plants** generating steam, and an **Electrical substation** generating electricity for the operations in the plant, **Wastewater collecting and treating** systems to make used water suitable for reuse or disposal.

The output from these processes are classified into three main classes of hydrocarbons based on their molecular structure.

Saturated hydrocarbons: These include two groups: paraffins or alkanes (this group includes methane, ethane, propane, butane, pentane, hexane, heptane, octane, nonane, and decane), and naphthanes or cycloalkanes.

Fig. 2.3 Oil refinery Silhouette. Figure from https://pixabay.com/images/search/oil%20refinery/

Fig. 2.4 Exxon Mobil oil refinery in Baton Rouge, Louisiana (the 4th largest in the United States), from reference [2]

Unsaturated hydrocarbons or Olefins: These include two groups: alkenes, and alkynes.

Aromatic hydrocarbons: These compounds are important petrochemical feedstocks, and in gasoline they help increase the octane number.

Gaseous fuel outputs like propane are stored and shipped in specialized vessels at high pressure to maintain them in liquid form. Lubricants (light machine oils, motor oils, greases, etc.) are usually shipped in bulk to an offsite packing unit. Paraffine wax is usually shipped in bulk to another site for preparing it in the form of packaged blocks. Sulfur, sulfuric acid, and byproducts of sulfur removal from petroleum are sold as industrial materials. Bulk tar is sent to an offsite unit for packaging it for use in tar-and-gravel roofing and other uses. Asphalt produced in various stages (Fractionation Towers in the CDU, Vacuum Distillation Unit, Fluid Catalytic Cracker Unit) is shipped in bulk to customers. Petroleum Coke is either sold as solid fuel, or to makers of speciality carbon products. Other liquid fuel finished products are stored in tanks allocated for their storage, and then used in blending them into gasoline, diesel, kerosene, aviation fuel, etc.

The industry is usually divided into three major sectors: upstream, midstream, and downstream.

Fig. 2.5 Oil refinery in Iran, from reference [2]

Fig. 2.6 Anacortes refinery (Marathon), on the north end of March point southeast of Anacortes, Washington, United States, from reference [2]

Fig. 2.7 Esso Oil Refinery at Fawley on Southampton Water. Opened in 1951, the refinery covers 3,250 acres of land, contains more than 330 giant tanks, has towers taller than football pitches, employs 1400 people regularly with 900 permanent staff, spends over £65 million every year in the local community. Photo taken from Weston Shore, Southampton. Figure from https://commons.wikimedia.org

The **upstream sector** includes exploration, development, and production of crude oil, natural gas.

The **midstream sector** operations and processes for crude oil include gathering (this operation employs narrow low-pressure pipelines to connect oil and gas producing wells to larger long-haul pipelines or processing facilities), refining, processing (this includes all the operations in the oil refinery to turn crude oil into products like heating oil, gasoline for road vehicles, diesel oil, and jet fuel) that can be marketed (these operations include distillation, catalytic reforming, catalytic cracking, alkylation, isomerization, and hydrotreating), transportation to processing facilities and then to end users by pipeline, tanker/barge, truck, and rail; storage (storage facilities are located near refining and processing facilities, and are connected by pipeline systems to facilitate shipment).

The **downstream sector** includes oil tankers, retailers, and consumers.

Immense Benefits to Humanity from Crude Oil

Detergents, soaps, and shampoos are made from crude oil derivatives; all these products improve sanitization, and make it possible for us to maintain good personal hygiene and good health.

A wholesome diet is a fundamental requirement for good health. Many goods including fresh vegetables, variety of fruit and other food items, also consumer products, equipment and machinery are transported from where they are produced to

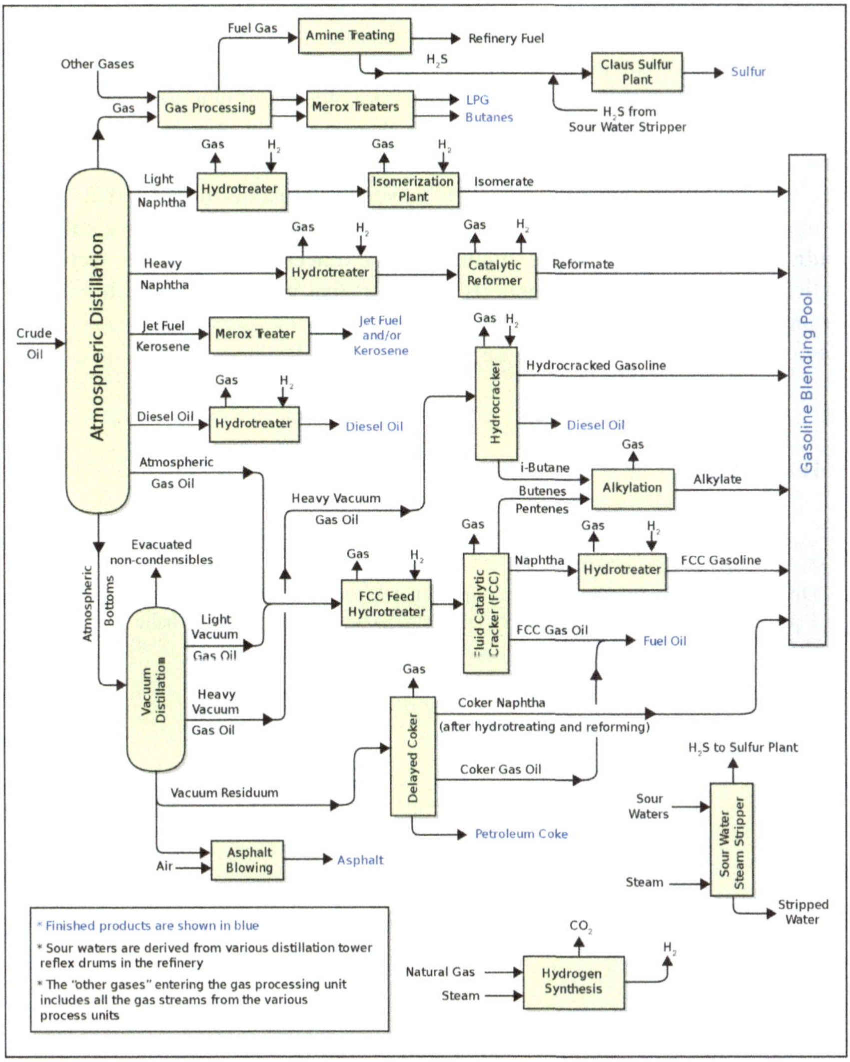

Fig. 2.8 Schematic flow diagram of a typical oil refinery, from reference [2]

where they are consumed; all this freight transportation is made possible, thanks to fuel derived from crude oil.

All the material used to make all the equipment that we use in our daily living may involve mining, processing, production, assembling, and marketing; all these make use of a lot of fuel derived from crude oil.

There is nothing on the horizon that has the energy density of gasoline and diesel, that is why demand for crude oil continues to go up yearly. But burning gasoline, diesel, fuel oil, all products made from crude oil, releases CO_2 and other **Green House Gases (GHG)** into the atmosphere. Growing accumulation of GHG in the atmosphere is believed to be the prime cause of **Climate Change** that we are seeing today. With the aim of reducing GHG gas emissions into the atmosphere, they have developed electric vehicles (road vehicles that run on electricity from a battery stored in the vehicle, and they are installing solar photovoltaic cells that produce electricity from sunlight, and windmills that produce electricity from the energy in blowing wind; all the produced electricity will be stored in huge batteries). At the moment it is not clear what percentage of road vehicle power requirements these alternate energy sources will meet, but even if this percentage reaches close to 100, **we still need crude oil for all the chemicals (including pharmaceuticals, solvents, fertilizers, pesticides, plastics, and synthetic fragrances)** that we make from it to maintain our present lifestyle.

For more oil refinery pictures see the link for "Oil Refinery Photos" on Google.

References

1. "Case Studies in Operations Research: Applications of Optimum Decision Making", Katta G. Murty (editor), Springer, 2015.
2. "Oil_ Refinery", from Wikipedia, see https://en.wikipedia.org/wiki/Oil_refinery
3. "Planning and Integration of Refinery and Petrochemical Operations", Khalid Y. Al-Qahtani, Ali Elkamel, Wiley-VCH, 2010.

Chapter 3
How Much Crude Oil Should an Oil Exporting Country Produce Annually for Export?

Katta G. Murty

Abstract In 1985 crude oil price was at \$31.75/barrel; and Saudi Arabia increased their production by 3% , hoping to increase their revenue from crude oil exports; but instead in a few months the crude oil price bottomed out at \$9.75/barrel. We discuss the international political dealings that led to a repeat of a similar event in 2014–2015. For a crude oil exporting country, several objective functions play a role in making a decision on how much crude oil to produce annually for export. In this chapter we discuss them, and also a procedure to find the best solution.

Historical Background

At the beginning of the year 2015, international press had many headlines with news about the resurgence of the US Oil and Natural Gas industry using the newly developed horizontal drilling and hydraulic fracturing (fracking) techniques. These techniques of course require huge investments. News reports indicated that with these investments US natural gas production will go up so much that the country won't have to import natural gas; and even the US crude oil production will go up to such an extent that pretty soon the country will move from an oil importing country to an oil exporting country. Crude oil produced using these new techniques is usually referred to as **shale oil**. For a background, we remind the reader that for many years, the US has been a net importer of crude oil with 2014 imports of 7 million barrels/day.

OPEC (Organization of the Petroleum Exporting Countries) is a cartel of 14 major oil exporters including Saudi Arabia, Iran, Iraq, Kuwait, and UAE accounting for one-third of global oil output. OPEC nations were producing 33.7 million barrels/day in 2015. **Saudi Arabia (SA)**, the leader of OPEC, is the world's largest exporter of

K. G. Murty (✉)
Department of IOE, University of Michigan, Ann Arbor, MI 48109-2117, USA
e-mail: murty@umich.edu

K. G. Murty (ed.), *Models for Optimum Decision Making*, International Series in Operations Research & Management Science 286, https://doi.org/10.1007/978-3-030-40212-9_3

crude oil; it exported 7.4 million barrels/day at the beginning of 2015. Russia which is not a member of OPEC is the second largest crude oil exporter. Both SA and Russia mainly depend on the revenue from crude oil exports for their sustenance. So, these countries, as well as all other oil exporting countries were mainly interested in raising the crude oil price to maximize their revenue from its sale at that time.

At that time (2014) crude oil price was ranging between \$100 and 110/barrel (in this book the symbol "\$" denotes "US dollar"). Due to huge public expenditures in the country, SA's budget break-even crude oil price was estimated to be \$88/barrel at that time.

The Two Players Involved in the Decision-Making Problem that We Will Discuss

One of the players with great interest in the outcome of the decision-making problem that we will discuss is the **US Government**. By 1973 crude oil imports by US made up a quarter of crude oil consumption in the country, and US was becoming increasingly dependent on imported crude oil; and by 2006 crude oil price reached \$75/barrel. For several years before 2015, the value of the US total annual imports was running higher than that of its total annual exports, resulting in a net annual trade deficit of around \$500 billion. This has a negative effect on the growth of the US economy. US was the world's largest importer of crude oil until 2014, also crude oil imports were a large part of US imports (as early as 2006 US was consuming about 20.7 million barrels of crude oil/day, 60% of it imported). So, the US is very much interested in keeping the crude oil price as low as possible.

Also, there was another major event that happened in Europe around this time. There was a civil war in the Crimean Peninsula (or **Crimea**) that was part of Ukraine at that time. The vast majority of inhabitants of Crimea are Russian people speaking the Russian language, and they never liked being a part of Ukraine. At that time the Russian Army marched into Crimea, and annexed it as a part of Russia. Supporting the Ukraine Government, the US Government imposed some **sanctions** on Russia. When these sanctions had no effect on Russia, the US Government was looking for some other actions they could take for punishing Russia. Since Russia depends heavily on their revenues from the sale of crude oil, reducing crude oil price will result in a severe punishment to Russia.

The **second player** is the **SA Government** represented by their Oil Minister. The income from their exports of crude oil is the major source of revenue for the country; another source for them is the revenue from the **Haj pilgrimage**. SA imports most of their groceries and several other commodities using this income, also SA's population is growing at the annual rate of about 2%. Given these facts, clearly SA is interested in keeping the crude oil price at the then (year 2014) prevailing rate of between \$100 and 110/barrel, and perhaps growing it on par with the growth rate in the prices of other commodities. In fact for a very long time, SA has been very modest in their desire for higher crude oil prices.

Other crude oil exporting countries have been trying vigorously to push up crude oil prices; as the world's largest producer of crude oil, SA has been playing the role of a moderator of these ambitions by increasing their crude oil production rate to keep price raises in check. The world really appreciates this moderating role that SA has been playing regarding crude oil prices.

History of Saudi Oil Industry

The **Kingdom of Saudi Arabia** was formed in 1932 with Abdulaziz Al-Saud as its King. A major source of income for the country at that time was the taxes paid by Haj Pilgrims to visit the holy sites of Islam. But the great depression of the 1930s resulted in the number of pilgrims/year decreasing from 100,000 to below 40,000. This hurt the Saudi economy greatly, so they started looking for an alternate source of income. Oil seepages have been observed at Quatif on the eastern seaboard, so the King invited mining engineers to explore the eastern regions of the country for oil. They identified a promising site near Dammam and over the following 3 years they started drilling for oil in that region. They stuck oil on March 3, 1938. This discovery turned out to be the first of many, eventually revealing the largest source of crude oil in the world. For the kingdom, oil revenues became a crucial source of wealth, and these discoveries altered Middle Eastern political relations forever. Soon the name of the company dealing with these oil producing operations was changed to **Arabian American Oil Company (Aramco)**. By 1988, Aramco was officially bought by the SA Government and became known as the Saudi Aramco.

Now SA is the world's largest producer and exporter of crude oil and has 25% of world's known oil reserves (over 264 billion barrels). As the world's largest producer and exporter of crude oil, SA plays the leading role in the global energy industry; its policies have a major impact on the energy market and the global economy. Mindful of this responsibility, the country is committed to ensuring stability of supplies and prices; so far they have been covering any drop in oil supplies by increasing their output.

With domestic demand and population raising rapidly, a very important objective for the SA Oil Minister to consider in reaching important decisions about oil production levels is to make sure that his country's domestic oil needs, and needs for revenues from oil exports, will be covered for *as long a planning horizon in future as possible*. This important objective function is measured by

Objective 1: Lifetime of the country's crude oil reserves resource = (country's estimated crude oil reserves)/(country's annual crude oil production rate) in years.

For Saudi Arabia the estimated oil reserves in 2015 were 264 billion barrels, and the country was producing 9.6 million barrels of crude oil daily at that time, so Objective 1 at that time was (264 billion)/(365 $\times$ 9.6 million) $\approx$ 75 years. After all, the oil reserves are a fixed finite resource that is being depleted by oil production. The SA Oil Minister is aware that about 40 years ago, the countries in Northern Europe were plush with oil from the reserves in the North sea; now these reserves are nearing depletion, and many of these countries are importing crude oil from Russia. One of his prime responsibilities is to make sure that this does not happen to his country SA, for the largest possible time interval in future, i.e., maximize Objective 1.

At that time they were exporting 7.4 million barrels/day of crude oil. At $100/barrel, this was providing oil revenues of $\$100 \times 7.4 \times 365$ million ≈ 270 billion/year. This revenue of $270 billion was meeting their needs for oil revenues at the beginning of 2015. This is the second important objective function to consider in our decision-making problem to decide how much crude oil the country should export annually, it is

Objective 2: Amount in US $ the country's exports of crude oil generate annually.

For Saudi Arabia at the beginning of 2015 it was $270 billion.

The third important objective function to consider in our decision-making problem to decide how much crude oil the country should export annually is:

Objective 3: Market share of the country in the oil export market = (the country's annual export of crude oil in barrels)/(Worldwide annual crude oil exports in barrels).

Saudi Arabia at the beginning of 2015 was exporting 7.4 million barrels/day, and the worldwide crude oil exports was 44 million barrels/day. So this Objective 3 had a value of 7.4/44 = 17% of market share in the crude oil exports market for Saudi Arabia at that time.

The Decision-Making Problem, Background in Early 2015

In early 2015, the US Government was looking for a way to reduce the price of crude oil for reasons mentioned at the beginning of the section "The Two Players Involved in the Decision-Making Problem that We Will Discuss". They already knew that in 1985 crude oil price was at $31.75/barrel, and Saudi Arabia decided to increase their crude oil production (thereby increasing world crude oil supply by merely 3%) with the hope of increasing their revenue from crude oil exports, but instead, in a few months the crude oil price bottomed out at $9.75/barrel [2]. From this historical fact, the US Government knew that they can achieve their goal of reducing the price of crude oil by persuading Saudi Arabia to increase their crude oil production. Fortunately for the US Government, at that time shale oil production in the US was beginning to take off. International News headlines were predicting that with shale oil production, US may change from an oil importing country into an oil exporting country. Mentioning this as a threat to Saudi Arabia's Market Share, **US Secretary of State** had discussions with the **Saudi Oil Minister** and other top level persons in the **Saudi Arabian (SA)** Government. He briefed them about these shale oil developments.

One of the countries in the Middle East with which SA does not have good relations is another OPEC member, **Iran**. At that time, the permanent members of the UN Security Council had imposed sanctions on Iran under the suspicion that the country is engaged in secret research to produce nuclear weapons, even though Iran has always denied these allegations. For many years before 2015, negotiations have been going on between Iran and these big six powers on conditions that Iran has to meet, to lift the sanctions placed on it. In discussions with Saudi authorities, US Secretary of State has pointed out that these Iran nuclear negotiations may at last be successful, and if so the sanctions on Iran will be lifted at that time. If that happens, Iran will begin oil production and enter the oil export market. He pointed out that these recent developments may affect **SA's market share in the crude oil export market**; and that they have to do something **to protect their market share**.

News reports in Western Media at that time indicated that the SA Oil Minister considered the emerging US Shale oil industry a big threat to the Saudi market share in the crude oil export market. So, in order to discourage additional investment in the US Shale oil industry, he decided to do everything in his power to make these investments unattractive. Knowing that this can be achieved by reducing the price of crude oil in the oil market, which can be achieved by flooding the market with crude oil, he took the decision of increasing the value of the country's crude oil production to the maximum extent that the Saudi oil industry can pump, and increase the country's crude oil exports with that extra production.

This recent Saudi action is contrary to the actions of all the OPEC members including SA until recently. Previously all the OPEC countries had the united policy of maintaining capacity limits for oil production by each country, to either increase the crude oil price or at least keep it from falling. That was to make sure that they are **rewarded adequately for making their precious natural resource available to the world**.

This action by SA was not popular with other OPEC members; they kept their crude oil production levels more or less stable. These actions raised the Saudi crude oil production to 10.6 million barrels/day, a record high. They resulted in the overall OPEC production increasing from 32 million barrels/day in 2014 to 33.1 million barrels/day in early 2015. The net result was that within a few days crude oil price dropped to 52$/barrel, and kept falling still further, a 50% drop due to the this action. That plunge from above 100$/barrel to 50$/barrel resulted in oil producing countries **losing billions of dollars in revenue**.

We will analyze this action taken by the Saudi Government in the next section.

Analysis of the Decision Taken

First, let us examine the consequences of the decisions implemented. In a few days after the decisions mentioned were implemented, the price of crude oil dropped by about 50%. So as a result of this action by SA, all the oil exporting countries experienced a similar decline in their returns/barrel exported. Prices of all other goods and commodities remained stable, so the net result was that SA is now having to export twice the amount of their precious natural resource to import the same quantity of other goods and commodities for their national needs, or use up some of their dollar reserves.

It is quite possible that maintaining market share is a matter of national pride, but is it worth having to pay twice the amount of the country's fixed natural resource for importing the same quantity of other goods and commodities? Also at the current production rate of 10.6 million barrels/day, the estimated lifetime of the current known reserves of SA will be (264 billion)/(10.6 $\times$ 365 million) $\approx$ 68.2 years, which is a significant drop of the other objective function, Objective 1, from its previous value of 75 years, as a result of the decision implemented.

The Procedure that OPEC IS Using Currently to Decide the Oil Production Quota of Each of Its Members

Each OPEC country wants to maximize the amount of money in US $ that they receive by exporting their crude oil, while making sure that their crude oil export volume is the same or higher than that in the recent past. OPEC's mission is to cooperate and unify the petroleum policies of its members and ensure the **stabilization of oil markets**. The Oil and Energy Ministers of OPEC members hold meetings at least twice a year to discuss global demand outlook and supply strategies and to coordinate their oil production policies. Each member country abides by an honor system in which everyone agrees to produce a certain amount. These production quotas of individual countries are agreed upon by mutual consent. A deal like this is not easy to reach, so OPEC constantly faces a big coordination problem. All members want to benefit from higher global prices, but that requires that their total production does not exceed global demand. Each country would rather see that someone else makes the necessary cuts in oil production. Some OPEC countries request minimal or no cuts in their oil production, or even permission to increase their oil production as their country needs to attract new investment to rebuild its sanctions ravaged industry (Iran), or from the ravages of war and the need for oil revenue to fight off ISIS (Iraq), or as their output was hurt by unrest and violence in the country (Libiya and Nigeria). Like this, OPEC constantly faces a big coordination problem. But in the end, in their meeting they reach a compromise.

Lessons to Be Learnt from the Events over the Last Two Years (2015–2017)

Crude oil prices have remained below US $50/barrel all this time, and even now with OPEC and other major crude oil producers like Russia limiting their crude oil production, crude oil prices are struggling to increase over 50 US $/barrel. Here briefly are the lessons we can learn from the events over this period.

1. To maintain the desired price level, unity of OPEC countries in abiding by agreed upon production limits is essential. Even if one country produces above their limit, the results may slip.

2. Keeping Objective 3, Country's market share at a high level may be a matter of National pride for the time being; but Objective 1, Lifetime of the country's crude oil resource is very important for the well-being of the next generation in the country. So, Objective 1 is very important and should not be ignored in decision-making.

3. Just increasing the country's crude oil production is not enough to make more money in US $, it is also necessary to make sure that the increased production does not result in lowering crude oil prices.

4. For every country, the country's crude oil reserves is a fixed finite resource, and crude oil production by any technique including fracking, depletes that resource.

5. When OPEC countries and others like Russia stick to their agreed upon crude oil production limits, if the US tries to keep crude oil price artificially low, by flooding the market with crude oil produced through fracking, they will soon realize that this effort is depleting their crude oil reserves resource faster than necessary.

6. To keep crude prices at desired levels, once this level is reached, it is necessary to make sure that the agreed upon production limits leads to available crude oil in the market equal to the demand for that oil.

Simple Approach to Calculate Crude Oil Export Quota for Each OPEC Country

Under these conditions, we get the following simple approach for calculating the crude oil export quota for each OPEC country [1]:

1. Each Country Selects its Desired Value for the Lifetime of its Crude Oil Resource: For each country, calculate (its crude oil reserves in barrels)/(its current annual crude oil production level in barrels) = its current value for "Lifetime of its crude oil crude oil resource". Then let the country modify that value to their "desired lifetime for its crude oil resources".

This implies that the country's "desired annual crude oil production rate" = (its crude oil reserves in barrels)/(its desired lifetime of its crude oil resource). Denote this quantity (in barrels/year) by the symbol y_i for OPEC Country i. Also, let

a_i = crude oil in barrels/year that the country needs for its internal uses and for its oil refineries within the country.

So, $y_i - a_i$ = amount of its production in barrels/year available for its crude oil exports.

2. Determine x = the Amount of Crude Oil in Barrels that OPEC as a Group Should Export for Coming Year: = (total estimated world crude oil export annual demand in barrels) – (expected amount that non-OPEC countries are likely to export if they are an oil exporting country, next year).

3. Distribute the Quantity x Among OPEC Countries in Proportion to Their Desired Annual Crude Oil production amount available for export calculated above, $(y_i - a_i)$: There are two cases to consider here:

Case 1: If $x \geq (\sum_p ((y_p - a_p)$: over all OPEC countries p)), make the crude oil export quota for the year as $(y_i - a_i)$ for OPEC country i; and leave $x - \sum_p ((y_p - a_p)$: over all OPEC countries p) as portion of the crude oil demand for the year, x, for other non-OPEC countries to fill.

Case 2: $x < (\sum_p ((y_p - a_p)$: over all OPEC countries p)): In this case, for OPEC country i, make its crude oil export quota $= x \times (y_i - a_i)/(\sum_p ((y_p - a_p)$:over all OPEC countries p)).

Example: Kuwait's oil reserves in 2015 were estimated to be 104 billion barrels. In 2015 they were producing 2.7 million barrels of oil per day, and exporting 1.7 million barrels per day from this. So Kuwait's annual crude oil production was $2.7 \times 365 = 986$ million barrels in 2015. So in 2015 the estimated lifetime of Kuwait's oil resource was $104000/986 = 105.5$ years. In 2015 to make the lifetime of their oil resource $= 120$ years, they should have reduced their daily production rate to $104000/(120 \times 365) = 2.4$ million barrels/day.

Exercise: Brazil's oil production rate was 2.5 million barrels/day in 2016, and their crude oil resource was estimated to contain 13 billion barrels. What was the estimate of the lifetime of their resource at that time? If they want to change that to 20 years, to what figure should they have changed their daily crude oil production at that time?

Exercise: In 2017 the small island nation of Bahrain in the Persian Gulf just east of Saudi Arabia's coastal areas and desert was producing around 45000 barrels per day from its onshore oil field, and additional 160000 barrels per day from an offshore field it shares with Saudi Arabia. In that year its proved resources of crude oil from these fields have been estimated to be 125 million barrels.

What is the estimated lifetime of Bahrain's current oil reserves at this rate of production in 2017?

In April 2018 Bahrain announced the discovery of 80 billion barrels of shale oil, its largest oil find ever, in the offshore Khaleej al-Bahrain basin spanning some 2000 km^2 in shallow waters off the country's western coast. Experts predicted that this newly discovered oil field is expected to be in production within 5 years. When this happens, how much will the country's lifetime of its crude oil reserves change if they keep their crude oil production rate the same as in 2017? Also, if they want to keep the lifetime of their crude oil reserves around 100 years, how much can they increase their crude oil production per day?

References

1. *Challenging Aspects of MCDM*, Katta G. Murty, 2015, see: http://www-personal.umich.edu/~murty/MCDM3.pdf
2. "Oil on the Brain: Petroleum's Long, Strange Trip to your Tank", Lisa Margonelli, Broadway Books, NY, 2007.

Chapter 4
Clustering Problems in Offshore Drilling of Crude Oil Wells

David Kaufman, Katta G. Murty and Ahmed AlSaati

Abstract An important decision-making problem in offshore drilling for crude oil is that of allocating various target locations in the field to drill, to available drilling rigs. In this chapter, we discuss an approach for solving this problem optimally.

First a brief introduction to offshore oil exploration. With worldwide consumption of crude oil over 30 billion barrels/year, and natural gas over 3500 billion cubic meters/year, crude oil and natural gas production and processing is the major industry all over the world today.

In 2009 the total world crude oil reserves have been estimated to be 1342 billion barrels, with Saudi Arabia as the country with the second highest reserves of 267 billion barrels. Its crude oil reserves are Saudi Arabia's most valuable asset, and Saudi Aramco handling these reserves is possibly the world's most valuable company.

Now crude oil and natural gas have become the most valuable commodities for international trade, and for countries exporting these commodities, the **quantity of their reserves** of these commodities has become a **hallmark of their economic status**. For Governments of countries among these who have some sea shore, this has provided a great incentive to explore for crude oil and natural gas reserves beyond their land border in their offshore areas. This has led to the development of **offshore drilling**, which has now become a thriving multibillion dollar industry in several

D. Kaufman (✉)
Decision Sciences, College of Business, University of Michigan, Dearborn, MI 48126-2638, USA
e-mail: davidlk@umich.edu

K. G. Murty
Department of IOE, University of Michigan, Ann Arbor, MI 48109-2117, USA
e-mail: murty@umich.edu

A. AlSaati
Well Completion Operations and Production Engineering Department, Aramco, Dhahran 31311, Saudi Arabia
e-mail: ahmed.saati@aramco.com

K. G. Murty (ed.), *Models for Optimum Decision Making*, International Series in Operations Research & Management Science 286, https://doi.org/10.1007/978-3-030-40212-9_4

areas of the world. However offshore drilling is much more expensive compared to conventional drilling on land.

Developing an area for hydrocarbons production begins with a complex process of geological surveys and analysis, seismic exploration, sedimentary basin analysis, and reservoir characterization (mainly in terms of the porosity and permeability of geologic reservoir structures) to identify underground rocks enriched with hydrocarbons indicating the presence of a reservoir of these hydrocarbons underground. Once such an area is identified, step-out wells drilled from mobile drilling rigs are used to determine the size and other characteristics of the field, and establish a reservoir model and determine the locations of **sweet spots** or suitable sites for production wells. By means of reservoir simulations, estimates of the total reserves and production as a function of time for a well at each of those sites, are obtained. A reservoir simulation is a computer model of underground rock and fluid conditions, which helps predict oil and gas recovery profile from a well at a target location. Of course this prediction depends on many factors like reservoir characteristics, flow properties, size, pressure near the well site, and the number of other wells already producing from the reservoir.

The information obtained from this activity is used to decide the locations for drilling of production wells, or **targets**. A typical 3×3 mile area of an offshore field will have between 25 and 300 production wells.

In offshore crude oil production, drilling of production wells is carried out by **drilling rigs** from either **fixed platforms** positioned on the seabed, or **subsea templates** placed on the seabed with flowline to a processing platform (for descriptions of these see Chap. 13 in Ref. [1]), or **bottom jackup rigs**, or **combined drilling and production facilities** with either **bottom grounded** or **floating platforms**, or **deep-water mobile offshore units** including semi-submersibles and drill ships. The choice of technology depends on parameters like the size of the field, water depth, and bottom conditions. In relatively shallow waters like the Gulf of Arabia (water depth upto 390 feet), a **jackup rig** (a self-elevating drilling platform with attached legs that can be lowered into the ocean floor and the hull raised above the sea level well above ocean waves and offering a relatively steady and motion-free platform for drilling) is most suitable.

A jackup rig can also be moved from one location to another; however, moving takes time and money, as all the equipment on top of the platform has to be taken down for this mobilization.

Drilling an oil/gas well is carried out in phases. At the end of each phase, a steel tube called **casing** is placed and **cemented inside the drilled hole**, and the next phase begins. As a result, each well drilled will be completed with several cemented tubulars inside each other all the way to the targeted depth. Once drilling the well is completed, a **wellhead** is installed in it on top of the sea floor where it will be connected to the production pipeline (Fig. 4.1).

Improvements in drilling technology have led to several improvements in drilling practices. Today we are not limited to drilling a well vertically all the way to the bottom. With **directional drilling**, it is now possible to drill deviated up to a horizontal direction. Directional drilling allows several oil producer wellheads to be

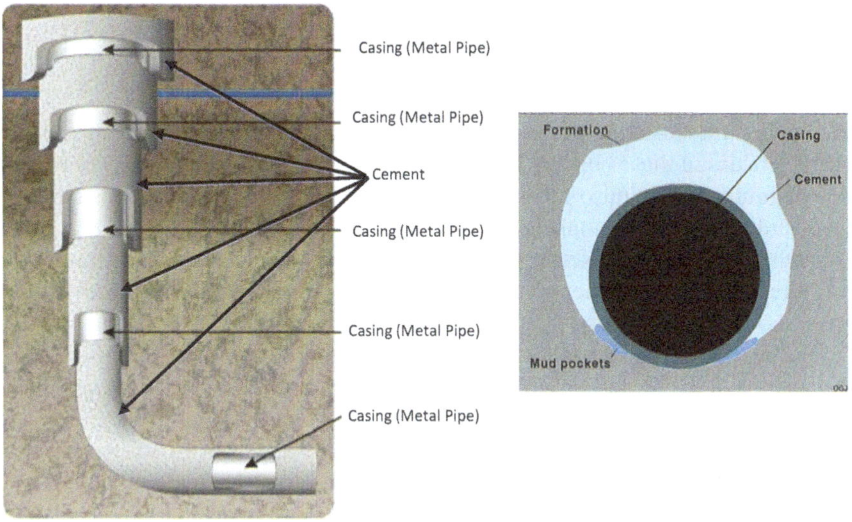

Fig. 4.1 Casing installed in the hole drilled, through which the crude oil being produced from the ground comes up. The outer layer around the casing is the cement layer that holds the casing in place

grouped together and all drilled with the jackup rig fixed in one location; and it allows increased exposure to the reservoir rock.

In directional drilling, drilling of a well begins in the vertical direction first, then at a point of specific depth known as the **kickoff point** the well inclination begins to increase. With this directional drilling practice, it is not necessary to move the rig to each well location to drill it; it is quite possible to group several sweet spots where wells have to be drilled into a group and drill all of them one after the other with the rig fixed in a central location with all of them around that location.

Environmental Damage Caused by Offshore Drilling: Offshore drilling poses the damage of causing serious pollution of ocean waters from **leaking large quantities of crude oil into the ocean**, thereby harming the environment which is the home of numerous fish species, whales, and other sea creatures. For example, in September 2004 hurricane Ivan tore through the Gulf of Mexico; this weather phenomenon caused the Taylor Energy Company's oil rig platform to collapse at the ocean bottom. The National Oceanic and Atmospheric Administration and Florida State University estimate that this collapsed oil rig had been leaking around 4500 gallons of crude oil/day ever since. Even after spending over $ 666 million, the company has not been able to stop the spill from continuing the leak into the Gulf. See: https://www.dailykos.com/stories/2019/6/26/1867500/-Oil-company-has-been-leaking-4500-gallons-of-oil-per-day-into-the-Gulf-of-Mexico-for-15-years?detail=emaildkre.

Application Involving Clustering Problems in Offshore Drilling

The input for a clustering problem is a set of objects, with relevent data about some characteristics of these objects. The desired output is a partition of the set of objects into **disjoint clusters** (also called classes or groups) satisfying certain constraints on their cardinality, that minimizes a specified objective function [2].

In this chapter, we discuss an important application of **clustering models** in offshore oil field development. This section is based on work carried out in developing an offshore oil field located 60 miles off the Eastern coast of Saudi Arabia.

The cost of drilling a production well depends on the distance of the well from the position of the rig and several other factors. Getting this cost exactly is very hard; it is estimated approximately using information from past drilling jobs.

Given the number and locations of production well targets in an offshore oil field to be developed, the problem is to partition the set of production well targets into clusters or groups, each group to be drilled by a single rig fixed in a single location; and also determine the best location to position the rig, to minimize the total cost of developing the field. This is a clustering problem involving huge sums of money. We discuss the mathematical model used to solve the problem. One parameter that is used for this decision-making is

$D = 10{,}500$ feet, maximum value of $y =$ distance between the target location to be drilled, and the current position of the rig, for the rig to drill this target without moving.

If $y > D$, then drilling the long lateral position in the horizontal direction is itself more expensive than that of moving the rig to the target location, so this rig in its present position will not be considered for drilling that well. Also a large value of D may be undesirable for other reasons besides its cost. So, the value chosen for D may vary from country to country, hence in an application, find out the preferred value for it; here we illustrate with $D = 10{,}500$ feet.

When a rig fixed in a position has to drill several target locations around it, it typically drills only one **vertical hole**, and from that reaches all the target locations it has to drill by drilling **horizontal sections**. The crude oil being produced from all these target locations has to come up to the ground from the casing installed in that single vertical hole. So, the number of target locations that the rig in its present position can drill is limited by the capacity of that casing to accommodate all the crude oil being produced from these target locations.

Several different cost components contribute to the cost of drilling a distance of length ℓ feet (either horizontal or vertical) using a rig. These various components can be approximated by linear functions of ℓ, obtained by linear regressions based on past drilling data from a field in the Gulf of Arabia, but may vary from field to field.

(i) Daily working rates paid to the drilling contractor: In crude oil production, the daily rate is the amount a drilling contractor gets paid by the oil company, for operating his rig per day. In offshore drilling, the daily rates of drilling rigs vary by their capability, and the market availability at the time the contract for the rig is signed.

The number of days required to drill a well of length ℓ feet has been estimated to be $0.0147\ell + 10.081$

(ii) Cementing: Cement is used to hold casing in place and to prevent fluid migration between subsurface formations. The price of cementing jobs vary based on additives used in mixing, well depth, well geometry, and rock characteristics in the field. From previous wells drilled and completed in this field, cementing cost has been estimated to be $ 16,850/foot of length.

(iii) Drilling fluid: Drilling fluids, also referred to as drilling mud, are added to the wellbore to facilitate the drilling process by controlling pressure, stabilizing exposed rock, providing buoyancy, and cooling and lubricating while drilling. This also depends on the type of fluid used and rock characteristics. From previous wells completed in this field, this cost has been estimated to be $94,671/foot of length.

(iv) Transportation, overhead, and other services: These expenses refer to transportation of equipment, tools and personnel to and off the rig, spare parts and iron tubes used to run the completion; drilling companies overhead such as office, personnel, fuel, water, logistics, and third party vendor services. We will subdivide these costs into three main categories, these are

Cost of overhead: estimated to be $ 80,080 per foot of length of the well

Cost of transportation: estimated to be $ 118,170 per foot of length of the well

Cost of services: estimated to be $ 280,280 per foot of length of the well.

So, all these costs sum to 590051ℓ + (daily working rates paid to the drilling contractor for $(0.0147\ell + 10.081)$ days).

The total length of the vertical portions to be drilled depends on the number of vertical holes drilled (which depends on the choice of the parameter D defined above), and the depth of the holes to be drilled; which are not factors that we can choose; so in a sense they are fixed. But the total length of the horizontal portions to be drilled depends on the order in which the rigs drill the wells; the optimum order minimizing the sum of the horizontal portions to be drilled by all the rigs can be found using the following procedure.

Let n be the total number of target locations to be drilled, and m the total number of rigs available. Also, let r_k be the drilling rate of the kth rig in feet drilled per hour, $k = 1$ to m.

Let $\mathcal{P}$ be the **shortest Hamiltonian path (SHP)** covering all the n target locations to be drilled, with the distance between each pair of these target locations as the cost coefficient corresponding to that pair; i.e., the path of minimum total length containing all these target locations (Fig. 4.2).

Divide $\mathcal{P}$ into m **segments** with the kth segment $\mathcal{P}_k$ containing the nearest integer to $n(r_k/(r_1 + r_2 + \ldots + r_m))$ of the n target locations. Then allocate the target locations on the kth segment, $\mathcal{P}_k$ as those to be drilled by the kth rig, $k = 1$ to m.

Here we defined the division of the shortest Hamiltonian path into segments containing target locations in proportion to the drilling rates of the available rigs. Instead, it is also possible to form segments of lengths proportional to the drilling rates of available rigs.

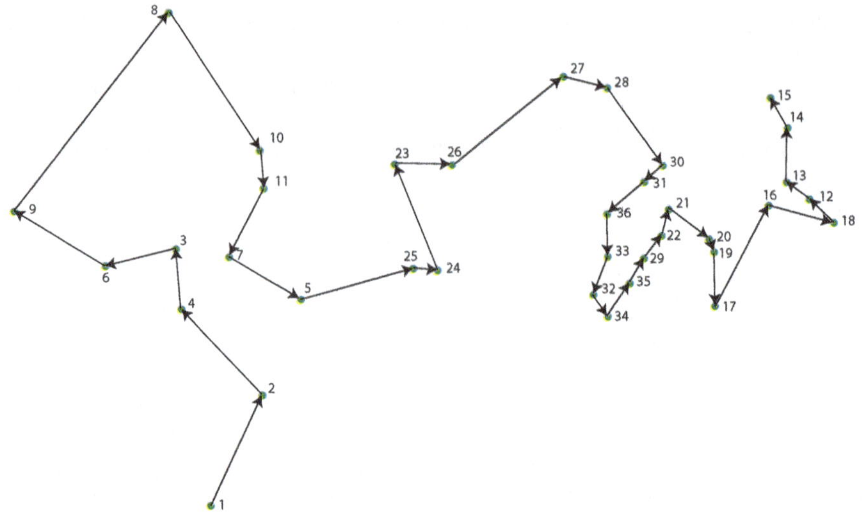

Fig. 4.2 Shortest Hamiltonian path $\mathcal{P}$ connecting all 36 target locations. The figure has been distorted in order to mask the coordinate data

Then the following idea is used repeatedly to generate a plan for this drilling: consider three consecutive target locations $i, i+1, i+2$ on $\mathcal{P}_k$. After drilling target location i, suppose the drilling bit of rig k drills the next target location $i+1$ on this segment by horizontal drilling along the edge $(i, i+1)$. In this situation, if the angle between the edges $(i, i+1)$, $(i+1, i+2)$ is not too much of a bend, then after horizontal drilling to target location $i+1$, that drilling bit can continue horizontal drilling along the edge $(i+1, i+2)$ to target location $i+2$.

To use this idea, we can identify a sequence of consecutive edges on $\mathcal{P}_k$ with the property that no pair of consecutive edges in it have too much of a bend (i.e., that the drilling arm can continue drilling the second edge in this pair after drilling the first, without returning to the rig). We will refer to such a maximal sequence (i.e., a sequence to which no more edges can be added while keeping this property) of consecutive edges on $\mathcal{P}_k$ as a **straight portion** of $\mathcal{P}_k$.

So, given a *straight portion* of $\mathcal{P}_k$, a rig at one end of this portion can send its drilling arm to horizontal-drill all the way to the last target location on it, subject to the horizontal distance D limit.

If there are two *straight portions* of $\mathcal{P}_k$ with a common end node, then by moving the rig to that node its drilling bit can horizontal-drill one of these *straight portions*, return to the rig and then it can horizontal-drill the other *straight portion*.

For each of the remaining individual target locations, the rig can drill them one after the other with the drilling bit returning to the rig after each.

Software for solving the Shortest Hamiltonian Problem (SHP) is available for download. For example, consider an SHP covering s nodes with the matrix $c = (c_{ij})$ of order $s \times s$, where c_{ij} = cost of traveling from node i to node j, for $i, j = 1$

to s. This is equivalent to the **Traveling Salesman Problem (TSP)** with C of order $(s+1) \times (s+1)$ as the cost matrix, where nodes 1 to s are the same as the nodes in the above SHP, and node $s+1$ is an **artificial node** with both the costs of traveling to and from node $s+1$ to and from every other node is 0. Given an optimum tour for this TSP, just deleting this artificial node and arcs incident at it, from this tour, leaves the shortest Hamiltonian path for our SHP problem. One software package for solving the TSP is the "Concorde TSP Code", the source code for which can be downloaded at

http://www.math.uwaterloo.ca/tsp/concorde/downloads/downloads.htm

Information on this **Concorde TSP solver** is available, for example, in Mathematica. My thanks to **William J. Cook, University of Waterloo**, for sharing this information with me.

Example: As an illustration, we discuss a problem with data taken from reference [3]. It discusses a problem in which there are $n = 36$ target locations numbered 1 to 36 to be drilled in a 15×8 mile area in the Persian Gulf, and there are $m = 4$ existing jacked up drilling rigs available for drilling these target locations.

Let c_{ij} denote the distance in miles between target locations i, j, for $i, j = 1$ to n. Let (i, j) denote the arc connecting target locations i, j. The matrix $C = (c_{ij})$ of these arc lengths is given in Tables 4.1, 4.2.

The shortest Hamiltonian path connecting all these 36 target locations has been obtained using the software mentioned above. In Table 4.3, we give the arcs in order along the Hamiltonian path together with the length in miles of each.

For explaining the conclusion from this solution, we illustrate this Hamiltonian path in the following figure. For drawing this figure, we are unable to use the actual coordinates of the target locations, due to the company's policy of not disclosing original data, so we based it on approximate data (Fig. 4.2).

In this problem, the drilling rates of all the four rigs available for drilling wells at these target locations are equal, i.e., all r_k have the same value. So, we divide this Hamiltonian path into four segments denoted by $\mathcal{P}_1, \mathcal{P}_2, \mathcal{P}_3, \mathcal{P}_4$ in that order, each segment containing 36/4 = 9 target locations. The sequence of target locations on these four segments are shown below (Fig. 4.3).

$\mathcal{P}_1$ is 1 - 2 - 4 - 3 - 6 - 9 - 8 - 10 - 11.
$\mathcal{P}_2$ is 7 - 5 - 25 - 24 - 23 - 26 - 27 - 28 - 30.
$\mathcal{P}_3$ is 31 - 36 - 33 - 32 - 34 - 35 - 29 - 22 - 21.
$\mathcal{P}_4$ is 20 - 19 - 17 - 16 - 18 - 12 - 13 - 14 - 15.

The bottom row in Table 4.3 corresponds to the edge connecting each segment to the next.

Figure 4.3 shows how the solution is implemented in practice. Assuming that for $k = 1$ to 4, rig k is assigned to drill the set of target locations in Segment k.

Segment 1 consists of target locations 1 to 11. It has 3 *straight portions*; $\mathcal{P}_{11}$ consisting of edges (1, 2), (2, 4), (4, 3) of length 1.57 miles; $\mathcal{P}_{12}$ consisting of edges (3, 6), (6, 9) of length 1.24 miles; $\mathcal{P}_{13}$ consisting of edges (11, 10), (10, 8) of length 1.33 miles.

Table 4.1 c_{ij} values for $i = 1$ to 36, $j = 1$ to 18, in miles

	$j = 1$	2	3	4	5	6	7	8	9	10	11	12	13	14	15	16	17	18
$i = 1$	0	0.57	0.85	0.9	1.31	1.51	2.2	2.05	2.08	3.1	3	1.93	2	1.93	2.21	2.52	2.64	2.56
2	0.57	0	0.77	0.8	0.88	1.34	1.73	1.88	1.9	2.41	2.63	2.11	2.08	2	2.42	2.61	2.59	2.6
3	0.85	0.77	0	0.2	0.69	0.64	1.5	1.2	1.4	1.78	2.18	1.35	1.38	1.4	1.75	1.8	1.83	2
4	0.9	0.8	0.2	0	0.7	0.7	1.6	1.3	1.3	1.7	2.1	1.4	1.35	1.38	1.8	1.9	1.8	2.2
5	1.31	0.88	0.69	0.7	0	1.2	0.88	1.15	1.2	1.6	1.78	1.92	1.92	1.92	2.36	2.11	2.06	2.2
6	1.51	1.34	0.64	0.7	1.2	0	1.17	0.58	0.6	1.16	1.62	1.25	1.24	1.35	1.72	1.39	1.34	1.3
7	2.2	1.73	1.5	1.6	0.88	1.17	0	1.19	1.2	1.23	1.04	2.41	2.43	2.4	2.89	2.26	2.09	2.29
8	2.05	1.88	1.2	1.3	1.15	0.58	1.19	0	0.2	0.61	1.21	1.37	1.35	1.36	1.83	1.14	0.94	1.1
9	2.08	1.9	1.4	1.3	1.2	0.6	1.2	0.2	0	0.81	1.01	1.57	1.55	1.56	2.03	1.34	1.14	1.3
10	3.1	2.41	1.78	1.7	1.6	1.16	1.23	0.61	0.81	0	0.72	1.79	1.86	1.82	2.27	1.32	1.03	1.03
11	3	2.63	2.18	2.1	1.78	1.62	1.04	1.21	1.01	0.72	0	2.52	2.62	2.5	3	2.04	1.76	2.09
12	1.93	2.11	1.35	1.4	1.92	1.25	2.41	1.37	1.57	1.79	2.52	0	0.1	0.2	0.8	0.87	1.11	0.87
13	2	2.08	1.38	1.35	1.92	1.24	2.43	1.35	1.55	1.86	2.62	0.1	0	0.05	0.48	0.84	1.1	0.84
14	1.93	2	1.4	1.38	1.92	1.35	2.4	1.36	1.56	1.82	2.5	0.2	0.05	0	0.53	0.89	1.15	0.89
15	2.21	2.42	1.75	1.8	2.36	1.72	2.89	1.83	2.03	2.27	3	0.8	0.48	0.53	0	1.15	1.46	1.17
16	2.52	2.61	1.8	1.9	2.11	1.39	2.26	1.14	1.34	1.32	2.04	0.87	0.84	0.89	1.15	0	0.31	0.02
17	2.64	2.59	1.83	1.8	2.06	1.34	2.09	0.94	1.14	1.03	1.76	1.11	1.1	1.15	1.46	0.31	0	0.35
18	2.56	2.6	2	2.2	2.2	1.3	2.29	1.1	1.3	1.03	2.09	0.87	0.84	0.89	1.17	0.02	0.35	0
19	3.15	3.63	2.32	2.2	2.36	2.53	3.44	1.19	1.39	0.92	1.51	1.57	1.73	1.78	2.07	0.92	0.63	0.72
20	3.24	3.06	2.39	2.3	2.38	1.78	2.1	1.25	1.45	0.87	1.4	1.88	1.9	1.95	2.31	1.1	0.8	0.89
21	3.31	3.2	2.46	2.5	2.4	1.94	2.17	1.32	1.52	0.97	1.44	1.96	1.98	2.03	2.24	1.16	0.87	0.96
22	3.24	3.17	2.47	2.5	2.47	1.86	2.16	1.28	1.48	0.93	1.46	1.7	1.93	1.98	2.28	1.15	0.87	1.27
23	2.99	3.1	2.35	2.3	2.73	1.99	2.93	1.78	1.98	1.96	2.69	1.07	1.06	1.11	1.08	0.69	0.93	1.02
24	2.96	3.2	2.35	2.35	2.73	2.01	2.96	1.81	2.01	1.97	2.7	1.1	1.06	1.11	1.08	0.68	0.93	1.02

(continued)

Table 4.1 (continued)

	$j = 1$	2	3	4	5	6	7	8	9	10	11	12	13	14	15	16	17	18
25	3.06	3.14	2.38	2.38	0.74	1.99	2.87	1.71	1.91	1.83	2.56	1.19	1.17	1.22	1.27	0.63	0.82	0.91
26	3.48	3.62	2.87	2.87	3.27	2	3.43	2.27	2.47	2.38	3.08	1.55	1.56	1.61	1.46	1.17	1.37	1.46
27	3.46	3.54	2.77	2.7	3.09	2.36	3.15	2.02	2.22	2.05	2.73	1.58	1.59	1.64	1.61	0.99	1.09	1.18
28	3.65	3.73	2.95	2.98	3.24	2.52	3.26	2.15	2.35	2.13	2.78	1.79	1.79	1.84	1.82	1.15	1.22	1.31
29	4.03	4.01	3.24	3.25	3.43	2.7	3.22	2.23	2.43	1.99	2.51	2.32	2.29	2.34	2.44	1.5	1.42	1.51
30	3.97	4.18	3.28	3.28	3.55	2.86	3.54	2.45	2.65	2.38	2.99	2.22	2.1	2.15	2.11	1.46	1.51	1.6
31	4.53	4.64	3.87	3.87	4.22	3.48	4.22	3.12	3.32	3.06	3.66	2.61	2.61	2.66	2.53	2.09	2.19	2.28
32	4.76	4.75	3.9	3.8	4.17	3.49	3.99	3.03	3.23	2.79	3.29	2.93	2.92	2.97	3	2.23	2.18	2.27
33	4.7	4.6	4.2	3.8	4.2	3.5	3.8	3	3.2	2.6	3.2	3	2.8	2.85	2.8	2	2.1	2.19
34	4.5	4.75	4	4.1	4.2	3.49	4.1	3.1	3.3	2.85	3.4	2.93	3	3.05	3.2	2.5	1.25	1.34
35	4.6	4.56	3.82	3.82	3.89	3.25	3.63	2.72	2.92	2.42	2.85	2.88	2.91	2.96	3.04	2.09	1.96	2.05
36	4.91	5	4.23	4.23	4.54	3.8	4.49	3.41	3.61	3.29	3.84	3	3.04	3.09	2.97	2.43	2.49	2.58

Table 4.2 c_{ij} values for $i = 1$ to 36, $j = 19$ to 36 in miles

	$j = 19$	20	21	22	23	24	25	26	27	28	29	30	31	32	33	34	35	36
$i = 1$	3.15	3.24	3.31	3.24	2.99	2.96	3.06	3.48	3.46	3.65	4.03	3.97	4.53	4.76	4.7	4.5	4.6	4.91
2	3.63	3.06	3.2	3.17	3.1	3.2	3.14	3.62	3.54	3.73	4.01	4.18	4.64	4.75	4.6	4.75	4.56	5
3	2.32	2.39	2.46	2.47	2.35	2.35	2.38	2.87	2.77	2.95	3.24	3.28	3.87	3.9	4.2	4	3.82	4.23
4	2.2	2.3	2.5	2.5	2.3	2.35	2.38	2.87	2.7	2.98	3.25	3.28	3.87	3.8	3.8	4.1	3.82	4.23
5	2.36	2.38	2.4	2.47	2.73	2.73	0.74	3.27	3.09	3.24	3.43	3.55	4.22	4.17	4.2	4.2	3.89	4.54
6	2.53	1.78	1.94	1.86	1.99	2.01	1.99	2	2.36	2.52	2.7	2.86	3.48	3.49	3.5	3.49	3.25	3.8
7	3.44	2.1	2.17	2.16	2.93	2.96	2.87	3.43	3.15	3.26	3.22	3.54	4.22	3.99	3.8	4.1	3.63	4.49
8	1.19	1.25	1.32	1.28	1.78	1.81	1.71	2.27	2.02	2.15	2.23	2.45	3.12	3.03	3	3.1	2.72	3.41
9	1.39	1.45	1.52	1.48	1.98	2.01	1.91	2.47	2.22	2.35	2.43	2.65	3.32	3.23	3.2	3.3	2.92	3.61
10	0.92	0.87	0.97	0.93	1.96	1.97	1.83	2.38	2.05	2.13	1.99	2.38	3.06	2.79	2.6	2.85	2.42	3.29
11	1.51	1.4	1.44	1.46	2.69	2.7	2.56	3.08	2.73	2.78	2.51	2.99	3.66	3.29	3.2	3.4	2.85	3.84
12	1.57	1.88	1.96	1.7	1.07	1.1	1.19	1.55	1.58	1.79	2.32	2.22	2.61	2.93	3	2.93	2.88	3
13	1.73	1.9	1.98	1.93	1.06	1.06	1.17	1.56	1.59	1.79	2.29	2.1	2.61	2.92	2.8	3	2.91	3.04
14	1.78	1.95	2.03	1.98	1.11	1.11	1.22	1.61	1.64	1.84	2.34	2.15	2.66	2.97	2.85	3.05	2.96	3.09
15	2.07	2.31	2.24	2.28	1.08	1.08	1.27	1.46	1.61	1.82	2.44	2.11	2.53	3	2.8	3.2	3.04	2.97
16	0.92	1.1	1.16	1.15	0.69	0.68	0.63	1.17	0.99	1.15	1.5	1.46	2.09	2.23	2	2.5	2.09	2.43
17	0.63	0.8	0.87	0.87	0.93	0.93	0.82	1.37	1.09	1.22	1.42	1.51	2.19	2.18	2.1	1.25	1.96	2.49
18	0.72	0.89	0.96	1.27	1.02	1.02	0.91	1.46	1.18	1.31	1.51	1.6	2.28	2.27	2.19	1.34	2.05	2.58
19	0	0.19	0.24	1.78	0.52	1.34	1.16	1.63	1.25	1.27	1.09	1.48	2.16	1.89	1.8	2.1	1.55	2.37
20	0.19	0	0.1	0.25	1.51	1.55	1.33	1.8	1.42	1.42	1.14	1.6	2.26	1.93	1.85	2	1.55	2.46
21	0.24	0.1	0	0.1	1.57	1.55	1.36	1.79	1.4	1.4	1.08	1.57	2.23	1.86	1.8	2	1.47	2.4
22	1.78	0.25	0.1	0	1.56	1.6	1.31	1.78	1.39	1.4	1.08	1.76	2.22	1.86	1.7	1.86	1.47	2.41
23	0.52	1.51	1.57	1.56	0	0.1	0.22	0.51	0.54	0.75	1.38	1.15	1.55	1.91	2	1.91	1.97	1.94
24	1.34	1.55	1.55	1.6	0.1	0	0.23	0.52	0.54	0.75	1.38	1.03	1.56	1.89	2.1	1.95	1.95	1.94

(continued)

Table 4.2 (continued)

$j = 19$	20	21	22	23	24	25	26	27	28	29	30	31	32	33	34	35	36
25 1.16	1.33	1.36	1.31	0.22	0.23	0	0.56	0.39	0.59	1.17	0.91	1.49	1.74	1.9	2	1.74	1.87
26 1.63	1.8	1.79	1.78	0.51	0.52	0.56	0	0.4	0.52	1.26	0.72	1.08	1.58	1.65	1.7	1.76	1.48
27 1.25	1.42	1.4	1.39	0.54	0.54	0.39	0.4	0	0.21	0.9	0.51	1.11	1.36	1.5	1.4	1.44	1.46
28 1.27	1.42	1.4	1.4	0.75	0.75	0.59	0.52	0.21	0	0.73	0.31	0.96	1.16	1	1.2	1.28	1.29
29 1.09	1.14	1.08	1.08	1.38	1.38	1.17	1.26	0.9	0.73	0	0.68	1.23	0.81	1	1.2	0.59	1.34
30 1.48	1.6	1.57	1.76	1.15	1.03	0.91	0.72	0.51	0.31	0.68	0	0.68	0.87	0.9	0.85	1.07	0.97
31 2.16	2.26	2.23	2.22	1.55	1.56	1.49	1.08	1.11	0.96	1.23	0.68	0	0.87	0.84	0.88	1.34	0.41
32 1.89	1.93	1.86	1.86	1.91	1.89	1.74	1.58	1.36	1.16	0.81	0.87	0.87	0	0.06	0.1	0.56	0.74
33 1.8	1.85	1.8	1.7	2	2.1	1.9	1.65	1.5	1	1	0.9	0.84	0.06	0	0.19	0.6	0.8
34 2.1	2	2	1.86	1.91	1.95	2	1.7	1.4	1.2	1.2	0.85	0.88	0.1	0.19	0	0.6	0.9
35 1.55	1.55	1.47	1.47	1.97	1.95	1.74	1.76	1.44	1.28	0.59	1.07	1.34	0.56	0.6	0.6	0	1.27
36 2.37	2.46	2.4	2.41	1.94	1.94	1.87	1.48	1.46	1.29	1.34	0.97	0.41	0.74	0.8	0.9	1.27	0

Table 4.3 Shortest Hamiltonian path connecting all 36 target locations. Segments 1 to 4 on this path allocated to rigs 1 to 4 are the portions on this path from node 1 to node 11, node 7 to node 30, node 31 to node 21, and node 20 to node 15, respectively. Distances are in miles

From	To	Distance	From	To	Distance	From	To	Distance	From	To	Distance
1	2	0.57	7	5	0.88	31	36	0.41	20	19	0.19
2	4	0.80	5	25	0.74	36	33	0.80	19	17	0.63
4	3	0.20	25	24	0.23	33	32	0.60	17	16	0.31
3	6	0.64	24	23	0.10	32	34	0.10	16	18	0.02
6	9	0.60	23	26	0.51	34	35	0.60	18	12	0.87
9	8	0.20	26	27	0.40	35	29	0.59	12	13	0.10
8	10	0.61	27	28	0.21	29	22	1.08	13	14	0.05
10	11	0.72	28	30	0.31	22	21	0.10	14	15	0.53
11	7	1.04	30	31	0.68	21	20	0.10			

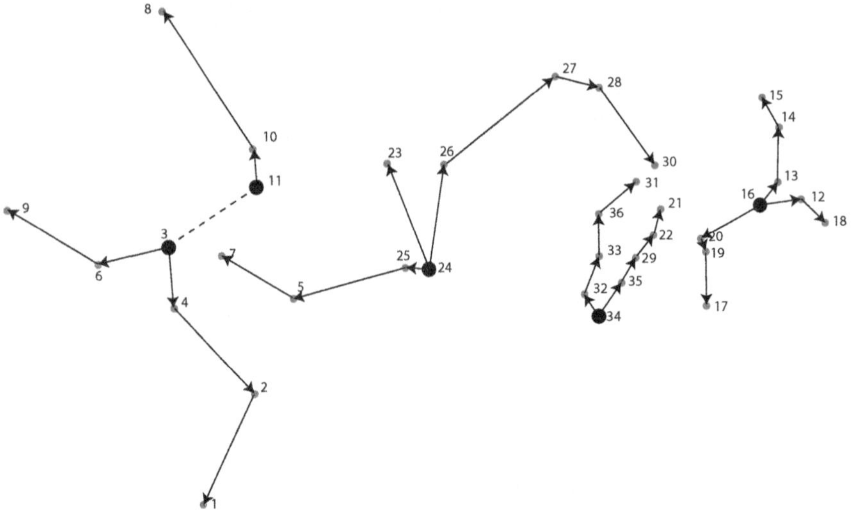

Fig. 4.3 Horizontal drilling segments for four rigs. (The figure has been distorted in order to mask the coordinate data.) Rig 1 starts at location 3 and drills two straight portions. Rig 1 is then moved (indicated by the dashed line) to location 11 where it drills one more straight portion. Rig 2 remains at location 24 and drills three straight portions. Rig 3 remains at location 34 and drills two straight portions. Rig 4 remains at location 16 and drills three straight portions

Rig 1 is first moved to target location 3; it drills a vertical hole there. Then its drilling arm drills locations 4, 2, 1 in that order through horizontal drilling. Then its drilling arm returns to the rig and from there it drills locations 6, 9 in that order and then returns to the rig at location 3. From the figure, we see that the pairs of lines {(3, 8), (8, 10)}, {(3, 10), (10, 11)} form too much of a bend. And the distance between locations 3 and 11 is 2.18 miles, so rig 1 is then moved to location 11 and it drills a vertical hole there. Then through horizontal drilling it drills locations 10, 8 in that order, and then returns to the rig at location 11. So, the total horizontal drilling carried out by Rig 1 is 4.14 miles, and this rig is moved once after its initial setup.

Segment 2 consists of target locations 7 to 30. It has 2 *straight portions*; $\mathcal{P}_{21}$ consisting of edges (24, 25), (25, 5), (5, 7) of length 1.85 miles; $\mathcal{P}_{22}$ consisting of edges (23, 26), (26, 27), (27, 28), (28, 30) of length 1.43 miles. Rig 2 is first moved to target location 24, and it drills a vertical hole there, then it drills target locations 25, 5, 7 in that order. Then through horizontal drilling, it drills locations 26, 27, 28, 30 in that order and returns to the rig at location 24 (total of 1.44 miles of horizontal drilling). Then its drilling arm carries out horizontal drilling to target location 23 (involving horizontal drilling of 0.1 miles), and finally returns to Rig 2 at its initial location of 24. So, the total horizontal drilling carried out by Rig 2 is 2.97 miles.

Segment 3 consists of target locations 31 to 21. It has two straight portions with a common end node, target location 34. $\mathcal{P}_{31}$ consisting of edges (34, 32), (32, 33), (33, 36), (36, 31) of length 1.37 miles; $\mathcal{P}_{32}$ consisting of edges (34, 35), (35, 29), (29, 22), (22, 21) of length 2.37 miles. Rig 3 is first moved to target location 34; it drills a vertical hole there, then it drills target locations 32, 33, 36, 31 in that order through horizontal drilling. Then its drilling arm comes to Rig 3, and from there, through horizontal drilling it drills target locations 35, 29, 22, 21 in that order; and finally returns to Rig 3 at its initial location 34. So the total horizontal drilling carried out by Rig 3 is 3.74 miles.

Segment 4 consists of target locations 20 to 15. It has 3 straight portions. $\mathcal{P}_{41}$ consisting of edges (15, 14), (14, 13), (13, 12), (12, 18) of length 0.75 miles; $\mathcal{P}_{42}$ consisting of edges (20, 19), (19, 17) of length 0.82 miles; $\mathcal{P}_{4,3}$ consisting of edges (17, 16), (16, 18) of length 0.33 miles. Rig 4 is first moved to target location 16; it drills a vertical hole there. Then through horizontal drilling, it drills target locations 20, 19, 17 in that order. Then its drilling arm returns to the rig location, and through horizontal drilling drills target locations 13, 14, 15 in that order and returns to the rig location. Then it horizontal-drills 12, 18 in that order, and finally returns to Rig 4 at its initial location 16. So the total horizontal drilling carried out by Rig 4 is 5.08 miles.

So, over all the rigs, we had to horizontal drill a total of 15.93 miles, and move one rig once after initial setup.

Exercise: Assume that the drilling rates of rigs 1 to 4 are proportional to (1.1, 1.2, 1, 1) in that order in the above example, but all the other data remains the same. What is the optimum solution in this case?

Exercise: In the above example if only three drilling rigs with equal drilling rates are available, but all the other data remains the same, what is the optimum solution?

Exercise: Consider the problem in which there are 30 target locations to be drilled at locations whose coordinates are {(0, 0), (1, 1), (1.5, 2), (4, 0), (4, 1), (7, 3), (7, 2.5), (7.5, 1.5), (8.7, 1), (9, 2), (10, 3), (11, 4), (11, 4.5), (9, 4), (7, 4.3), (6, 3.9), (5, 3.7), (4, 4), (3, 3), (3.1, 8), (2, 0.5), (3, 1), (4.2, 1), (4.5, 2), (5, 3), (5, 4), (4.8, 6), (4.9, 7), (4, 2.8),) (4.5, 10)}. The axes of coordinates are orthogonal with origin at the first target location, and distance along both axes is measured in units of (1/4)miles. There are three drilling rigs available with equal drilling rates. The parameter D in the problem is 2 miles as discussed above. Find the optimum allocation of target locations to drill, to drilling rigs, and how to implement it.

References

1. "Case Studies in Operations Research: Applications of Optimum Decision Making", Katta G. Murty (editor), Springer, 2015.
2. "Clustering Problems in Optimization Models", Santosh Kabadi, Katta G. Murty, Cosimo Spera; *Computational Economics*, 9, 229–239, Kluwer Academic Publishers, Netherlands, 1996.
3. *Clustering Models in Offshore Drilling for Crude Oil and Natural Gas*, Hesham K. Alfares, Katta G. Murty, Ahmed AlSaaty, Technical Report, King Fahd University of Petroleum and Minerals, Dhahran, Saudi Arabia, 2017.

Chapter 5
A Product Mix Optimization Application at the Saudi ARAMCO Company, Sadara Chemicals

Katta G. Murty

Abstract A crude oil producing country can either export the crude oil they produce directly, or set up oil refineries for refining the crude oil they produce and sell the products that this produces. In this chapter, from simple product mix examples, we show that the second alternative not only leads to much higher profits, but also provides excellent jobs for their countrymen.

Saudi Aramco has been the world's largest exporter of crude oil for a long time. In early 2000s they had the idea that instead of exporting crude oil directly, they can make more money by using that crude oil to manufacture and market high value-added chemical products and performance plastics used as chemicals and additives in the oil and gas industry; in making chemicals for water treatment and numerous consumer products like soaps, detergents, cosmetics and other personal care products, fibers, paints, bedsheets, automobile furnishings, adhesives and sealants, break fluids, and packaging materials; and in the energy, transportation, electronics, and furniture industries. In addition to higher revenues, this will also generate over 20,000 quality direct and indirect jobs for Saudi nationals, and also attract several downstream investments.

Of course the manufacture of all these new products will require lots of fresh water, of which Saudi Arabia does not have any natural supply. So if plants are set up to make these new products, their fresh water requirements have to be met using expensive desalinated sea water. The feeling is that in spite of this, these ventures will be profitable. We illustrate this with a small example.

Example: This example is based on an application in the late 1970s, and is based on prices and costs at that time. It is about a small refinery with a crude distillation capacity of 100,000 barrels of crude oil/day in its Crude Distillation Unit (CDU), and it has three different sources for crude oil. The crude supplies from these sources

---

K. G. Murty (✉)
Department of IOE, University of Michigan, Ann Arbor 48109-2117, USA
e-mail: murty@umich.edu

K. G. Murty (ed.), *Models for Optimum Decision Making*, International Series in Operations Research & Management Science 286, https://doi.org/10.1007/978-3-030-40212-9_5

will be referred to as crudes 1, 2, and 3, respectively. Their availabilities and prices are given below.

Crude availabilities and prices			
Data	Crude 1	Crude 2	Crude 3
Availability (barrels/day)	60,000	90,000	80,000
Price ($/barrel)	23.25	22.00	20.50

The main products that they use from distillation are Distillation Naphtha (DN), Distillation Heating Oil (DHO), Distillation Gas Oil (DGO), and Pitch (P). The estimated yields of these products from distillation of the different crudes in their CDU are:

Yield (barrels/barrel) from distillation of the 3 crudes			
Output	Crude 1	Crude 2	Crude 3
DN	0.19	0.16	0.02
DHO	0.27	0.32	0.24
DGO	0.38	0.27	0.26
P	0.05	0.13	0.39

The other product yields from distillation of these crudes are not considered in this example.

DHO can be sold directly as Heating Oil. DGO can be sold directly as Diesel oil. DHO, DGO can also be processed further in a catalytic cracker. The catalytic cracker can either process a maximum of 100,000 barrels/day of DHO, or a maximum of 50,000 barrels/day of DGO or a combination of these in proportion to these levels adding up to 1. Also, when processing DHO the catalytic cracker can be run either at normal level, or at a high Severity level with a higher production rate. DGO can be run at Normal level only, on the Catalytic Cracker. Data on the output levels on the Catalytic Cracker is given below: here CN is Catalytic Naphtha, CHO is Catalytic Heating Oil, and P is Pitch.

Outputs (barrels/barrel of feed) of products CN, CHO, P on catalytic cracker			
Feed and process	CN	CHO	P
DHO, Normal level	0.18	0.80	0.11
DHO, High Severity level	0.32	0.69	0.10
DGO, Normal level	0.48	0.52	0.15

Here are the marketable products that can be made by blending these outputs, with their selling prices and limits on demand.

Heavy fuel oil at $ 30/barrel: obtained by blending two parts of CHO with 17 parts of Pitch, unlimited demand.

Gasoline at \$ 43.25/barrel: obtained by blending 20 parts of DN with 17 parts or greater of CN. Quality of Gasoline improves with proportion of CN in this blend, demand up to 40,000 barrels/day.

Heating Oil at \$ 39.75/barrel: this is DHO sold directly, demand up to 40,000 barrels/day.

Diesel Fuel at \$ 39/barrel: this is DGO sold directly, unlimited demand.

The processing costs in \$ /barrel processed are \$ 0.6 for crude processed on the CDU, \$ 0.95 on catalytic cracker at Normal level, and \$ 1.50 on catalytic cracker at High Severity level.

Find an optimum plan that maximizes Net daily profit of this company.

The decision variables in this problem are listed below.

x_1, x_2, x_3: barrels/day of Crudes 1, 2, 3 processed on the CDU.

y_1, y_2, y_3, y_4: barrels/day of DN, DHO, DGO, and P, respectively, produced on the CDU.

y_{21}, y_{22}, y_{23}: DHO sold/day directly as Heating oil, processed on catalytic cracker at Normal, High Severity levels.

y_{31}, y_{32}: DGO sold/day directly as Diesel fuel, processed on catalytic cracker at Normal level.

u_1, u_2, u_3: barrels/day of CN, CHO, and Pitch produced on catalytic cracker.

v_1, v_2, v_3, v_4: barrels/day of Heavy fuel oil, Gasoline, Heating oil, Diesel fuel, sold.

v_{21}, v_{22}: barrels/day of DN, CN in the Gasoline blend.

The model for this problem is given below. In the model constraint (1) represents the distillation capacity of the CDU. Constraints (2)–(5) represent the yields for DN, DHO, DGO, and P at the CDU from the crude oils fed into it. Equations (6), (7) represent the three, two streams into which the DHO, DGO produced at CDU are split. Equation (8) represents the processing capacity of the catalytic cracker. Constraints (9)–(11) represent the yields for CN, CHO, and P at the catalytic cracker. Constraints [(12), (13)], {(14), (16)} represent the limits on the amounts on [Heavy fuel oil], {Gasoline} produced, imposed by the availability of constituents in the blending formulas for them. Equation (15) represents the constraint that the amount of CN used in the gasoline blend is less that the amount of CN produced in the catalytic cracker units. Constraints (17)–(19) represent the supply limits of the three suppliers of crude oil. Equations (20), (21) represent the blending of DN and CN to produce Gasoline, and 40,000 barrels/day is the upper limit for its demand in the market. Equations (22), (23) represent the two symbols used for Heating oil are equal, and 40,000 barrels/day is the upper limit for its market demand. Equation (24) represents that the two symbols used for Diesel Fuel are equal.

Maximized Net daily profit = $(30v_1 + 43.25v_2 + 39.75v_3 + 39v_4) - (23.25x_1 + 22.00x_2 + 20.50x_3) - 0.6(x_1 + x_2 + x_3) - 0.95(y_{22} + y_{32}) - 1.50y_{23}$
subject to:

$$x_1 + x_2 + x_3 \leq 100,000 \quad (1)$$
$$-y_1 + 0.19x_1 + 0.16x_2 + 0.02x_3 = 0 \quad (2)$$
$$-y_2 + 0.27x_1 + 0.32x_2 + 0.24x_3 = 0 \quad (3)$$
$$-y_3 + 0.38x_1 + 0.27x_2 + 0.26x_3 = 0 \quad (4)$$
$$-y_4 + 0.05x_1 + 0.13x_2 + 0.39x_3 = 0 \quad (5)$$
$$y_2 - y_{21} - y_{22} - y_{23} = 0 \quad (6)$$
$$y_3 - y_{31} - y_{32} = 0 \quad (7)$$
$$((y_{22} + y_{23})/100,000) + (y_{32}/50,000) \leq 1 \quad (8)$$
$$-u_1 + 0.18y_{22} + 0.32y_{23} + 0.48y_{32} = 0 \quad (9)$$
$$-u_2 + 0.80y_{22} + 0.69y_{23} + 0.52y_{32} = 0 \quad (10)$$
$$-u_3 + 0.11y_{22} + 0.10y_{23} + 0.15y_{32} = 0 \quad (11)$$
$$v_1(2/19) - u_2 \leq 0 \quad (12)$$
$$v_1(17/19) - u_3 - y_4 \leq 0 \quad (13)$$
$$v_{21} - y_1 \leq 0 \quad (14)$$
$$v_{22} - u_1 \leq 0 \quad (15)$$
$$v_2(17/37) - v_{22} \leq 0 \quad (16)$$
$$x_1 \leq 60,000 \quad (17)$$
$$x_2 \leq 90,000 \quad (18)$$
$$x_3 \leq 80,000 \quad (19)$$
$$v_2 - v_{21} - v_{22} = 0 \quad (20)$$
$$v_2 \leq 40,000 \quad (21)$$
$$v_3 - y_{21} = 0 \quad (22)$$
$$v_3 \leq 40,000 \quad (23)$$
$$v_4 - y_{31} = 0 \quad (24)$$

All variables are ≥ 0. (25)

The optimum solution of this LP is $(x_1, x_2, x_3) = (0, 20000, 80000)$, $(y_1, y_2, y_3, y_4) = (4800, 25600, 26200, 33800)$, $(y_{21}, y_{22}, y_{23}) = (25600, 0, 0)$, $(y_{31}, y_{32}) = (17700, 8500)$, $(u_1, u_2, u_3) = (4080, 4420, 1275)$, $(v_1, v_2, v_3, v_4) = (39201.47, 8880, 25600, 17700)$, $(v_{21}, v_{22}) = (4800, 4080)$, with the optimum objective value = Net daily profit = \$1.12 million.

In this example by selling crude oils used in the solution directly, we get \$1.64 million; but from the income we get by selling the products produced by processing that crude oil, after subtracting the price of crude oils used, and the costs of processing, we are left with a net profit of over \$1.12 million. This clearly shows that instead of selling crude oil directly, it is more profitable to process it and sell the products produced from this processing.

Exercise from this example (Geoffrey Gill): Let p_1 to p_{24} be the dual variables associated with constraints (1) to (24) in the above LP. In the dual problem, the dual variables $p_2, p_3, p_4, p_5, p_6, p_7, p_9, p_{10}, p_{11}, p_{20}, p_{22}, p_{24}$ are real variables unrestricted in sign, because they are associated with equality constraints in the primal; and the remaining dual variables are nonnegative variables. Write down the dual problem of this LP, and find its optimum solution. Using it answer the following questions:

(i): Crude 1 is not used in the optimum solution of this problem. Assuming that all the data in this example remains unchanged except the price of crude 1, determine at what price crude 1 will appear in an optimum solution of the problem.

(ii): Assuming that all the data in this example remains unchanged, how much capital expense can we incur to increase the capacity of the CDU and the catalytic cracker units to increase their capacity by 10,000 barrels/day?

Based on this preliminary analysis, Aramco has decided to go ahead with this venture and awarded the contract to Dow Chemical Co. to build these plants in Jubail, Saudi Arabia, and construction has commenced in 2007. By 2016, 26 manufacturing units have been set up with over 3 million tons capacity per year, and the construction was completed by 2018. This venture is expected to generate revenues of over \$ 10 billion per year. Exports will go to Asia (45%), Europe (10%), and Middle East (25%).

Also, encouraged by these results, Saudi Arabia, the world's largest crude oil exporter, made plans to invest in Asian refineries in China, India, Indonesia, Malaysia, and Vietnam; and build new refineries there as part of a plan to almost double its global refining capacity from around 5.4 million barrels a day in 2014 to around 10 million barrels per day globally in 10 years. It already exports crude oil to these countries, and by investing in refining crude oil there, they can ensure that they have plenty of buyers in the fastest growing region for fuel demand; and it will give the kingdom a key outlet for its crude oil in these countries, and gives it influence in determining pricing of oil products worth billions of dollars each day.

By the beginning of 2018, the construction of the Sadara Chemicals Complex in Jubail was completed and it is in operation at full capacity. Encouraged by the results, then Saudi Arabia took the decision to boost chemical output capacity worldwide to another 70% by 2025. They signed a multibillion dollar deal with the French oil and gas major Total to build another giant petrochemicals complex at their 440,000 barrels per day Jubail SATORP refinery which will create 8000 new jobs, and help take advantage of the fast growing Asian Polymer market. Also, Saudi Basic Industries Corporation, the Middle East's dominant chemical maker decided to work with new joint venture partners and expand its footprint in the heart of US Shale boom, and build a Houston Headquarters for its Western Hemisphere operations. And Aramco

Fig. 5.1 Figure shows a shell gas station, one of the end users of products from an oil refinery. Figure from https://pixabay.com/images/search/oil%20refinery/

took the first steps to integrating a petrochemicals business into the US biggest oil refinery at Port Arthur, Texas operated by their subsidiary Motiva Enterprises and boost its capacity to 1.5 million barrels per day, which will make it the largest in the world.

In 2018 Saudi Aramco took a 50% stake in a planned 1.2 million barrels/day $44 billion west coast mega refinery in India in a move that would give the kingdom a new stable outlet for its oil (Fig. 5.1).

Exercise: This exercise deals with the allocation of crude oil production to oil fields in Saudi Arabia. Crude oil coming up through an oil well contains various dissolved gas components. The Gas–Oil ratio (GOR) at a well measures the Cubic Feet (CF) of dissolved gases per barrel of gas–oil mixture coming up through the well; this differs from field to field. The gas–oil mixture coming up through the well is sent by a pipeline to a Gas Oil Separation Plant (GOSP) in which the gas and oil in the mixture are separated. The Shrinkage Factor (SF) of a well is the barrels of crude oil left after one barrel of gas–oil mixture coming up from the well is processed for gas removal in a GOSP, this differs from field to field.

From the GOSPs, the crude oil is transported to various refineries in the country, or to shipping terminals for export. The liberated gases are sent by pipeline to Gas Plants where sulfur (S) and methane (also called fuel gas) are separated from them and the remaining gas liquified into Liquified Natural Gas (LNG) is sent by pipeline to fractionation plants where it is separated into its various gas components: Ethane, Propane, Butane, Pentane, Naphtha, and others. Methane and Ethane are then piped to industrial complexes to be used either as fuel in electric power production, desalination plants, etc., or as feedstocks in chemicals manufacturing.

The GOR in the gas–oil mixture coming up through a well remains constant as long as the production is above the reservoir bubble point, which is the case in all the oil fields under consideration.

In this exercise we consider 4 oil fields F_1 to F_4, which have wells producing two types of crude oil: Arabian Light and Arabian Medium; 10 GOSPs labeled as G_1 to G_{10}; 3 Gas Plants, GP1, GP2, GP3; 2 Fractionation plants, FP1, FP2; and one LNG terminal from which LNG can be sent by pipeline to a fractionation plant, sometimes LNG can also be sent from a gas plant directly to a fractionation plant.

The crude oil produced at the oil fields F_1 to F_3 is Arabian Light; that produced at oil field F_4 is Arabian Medium.

Capacities for pipelines transporting liquids are measured in units of 10^5 barrels/day; and those for gases are measured in units of 10^6 CF (Cubic feet)/day. Oil field F_1 is connected by pipelines transporting liquids to GOSP G_1, G_2 of capacity 3 units each; F_2, F_3 are connected by pipelines transporting liquids to GOSPs G_3 to G_8 of capacity 2 units each; F_4 is connected by pipelines transporting liqiuids to GOSPs G_9, G_{10} of capacity 2.5 units each, respectively.

GOSPs G_1 to G_5 are connected to Gas Plant GP1 by pipelines for liquid transport of capacity 93 units; G_6, G_7, G_8 are connected to Gas Plant GP2 by pipelines for liquid transport of capacity 35 units; and G_9, G_{10} are connected to Gas Plant 3 GP3 by pipelines for liquid transport of capacity 19 units.

Crude oil separated at GOSPs from the crude oil–gas mixture is transported to various refineries in the country or to shipping terminals for export.

At each gas plant, the sulfur produced is sent to the market for it. From Gas Plant GP1 the LNG is sent by pipeline to Fraction plant FP1 for separation into the products Ethane, Propane, Butane, Pentane, Naphtha, and others. The LNG from Gas Plants GP2, GP3 is shipped by pipeline to an LNG Terminal which sends it by pipeline to Fraction plant FP2 for the same separation.

Each of the three gas plants has a capacity to handle 1500 units/day (unit is 10^6 cubic feet). Fractionation plants FP2, FP3 have capacities of 3, 7.3 units/day (unit is 100,000 barrels) of LNG.

Following tables give the data on the outputs from the four oil fields.

Cost (SR(Saudi Riyals)/barrel), GOR and SF at the four fields

Oil Field	Cost	GOR	SF
F_1	6.56	547	0.747
F_2	6.56	400	0.794
F_3	5.7	846	0.685
F_4	8.74	735	0.801

Processing costs at gas plants and fractionation plants are ignored but the total cost considered includes transportation costs. It costs 0.0034, 0.0114, 0.0021, 0.0021 SR/CF to transport the liberated gases from GOSPs 1 to 5, respectively, to Gas Plant 1; 0.156, 0.0165, 0.0312 SR/CF to transport them from GOSPs 6, 7, 8, respectively,

Outputs from a CF of gases at the four fields			
Oil Field	CF of methane	Grams of sulpher	Barrels of LNG
F_1	0.599	0.0167	0.0002
F_2	0.606	0.0461	0.00015
F_3	0.498	0.0254	0.0024
F_4	0.529	0.1105	0.0003

Output of components from 1 barrel of LNG					
Stream	Ethane in CF	Butane*	Pentane*	Propane*	Others
GP1-FP1	852.71	0.145	0.103	0.333	Rest
GP2- FP2	820.47	0.153	0.102	0.341	Rest
GP3-FP2	919.78	0.148	0.089	0.310	Rest

* = in barrels

to Gas Plant 2; and 0.00047 SR/CF to transport them from GOSP 9 or 10 to Gas Plant 3.

It costs 2.24 SR/barrel to transport LNG from Gas Plant 1 to Fractionation Plant 1. It costs 0.31. 0.127 SR/barrel to transport LNG from Gas Plants 2, 3 to the LNG Terminal. The cost of shipping LNG from the LNG Terminal to Fraction Plant 2 is 1.1 SR/barrel.

They would like to produce Arabian Light, Arabian Medium crude oils in proportion 60:40. For any deviation in the daily production of Arabian Light, Arabian Medium from this proportion, a penalty of 0.0025 SR/barrel is included in the overall cost.

The OPEC production ceiling would impose an upper bound of $(4.5)10^6$ barrels per day on the daily production of both types of crude oils put together from these four oil fields. They would like to adhere to this limit.

The internal demand for methane, ethane are $(1.621)10^8, (2.87)10^7$ CF/day respectively, which they would like to meet as far as possible. There is a penalty for shortage of either of these products at 1000 SR/CF.

Formulate the problem of determining how much gas–oil mixture to produce from each of the four fields under consideration, and the optimal routing of this production through the processes and plants described, so as to meet the demand for methane, ethane at minimum total cost, from [1].

Reference

1. "A Linear Programming Model to Evaluate Gas Availability for Vital Industries in Saudi Arabia", S.O. Duffuaa, J. A. Al-Zayer, M. A. Al-Marhoun, and M. A. Al-Saleh, Journal of the Operational Research Society, 43, no. 11 (1035-1045), Nov. 1992.

Chapter 6
Blending Operations in Crude Oil Refineries

Katta G. Murty and David Kaufman

Abstract Outputs from treatment units in an oil refinery are only semi-finished products. In this chapter, we discuss procedures for blending these products into finished products like gasoline, diesel oil, etc. meeting various specifications on them for selling in the marketplace.

The outputs from the CDU (Crude Distillation Unit), and other treatment units in an oil refinery are generally not commercially usable directly. They are only semi-finished products, which can be blended into products meeting various specifications that the marketplace demands. The problem of blending semi-finished products from the CDU and other processes into finished blends satisfying specifications that the marketplace demands on several of their properties leads to nonlinear models; but the industry has the desire to obtain reasonable approximations using simpler linear programming models. The focus of Chaps. 6 and 7 is to discuss techniques developed to satisfy this desire of the industry.

There are two classes of blending operations carried out in most refineries. The first is mixing different crude oils to prepare a crude blend with desired properties to feed into the CDU.

Reasons for Blending Crude Oils in Oil Refineries

The processing units in each crude oil refinery are designed to process certain slate of crudes based on their properties, a decision made depending on the availability and cost of crudes at the time the refinery was built. The more consistent the supply of crude oil to the specific refinery, the more that refinery can tailor its operation to that specific crude supply.

K. G. Murty (✉)
Department of IOE, University of Michigan, Ann Arbor 48109-2117, USA
e-mail: murty@umich.edu

D. Kaufman
Decision Sciences, College of Business, University of Michigan,
Dearborn 48126-2638, USA
e-mail: davidlk@umich.edu

K. G. Murty (ed.), *Models for Optimum Decision Making*, International Series in Operations Research & Management Science 286, https://doi.org/10.1007/978-3-030-40212-9_6

Fig. 6.1 Crude oils of different colors. The color of crude oils can vary between colorless or pale yellow to red, green, or most common dark brown to black. From left to right: Light to medium crude oils from the Caucasus, the Middle East, Arabia, and France. Figure from https://commons.wikimedia.org

In its natural, unrefined state crude oil ranges in density and consistency, from very thin, light weight, volatile fluidity; to a very thick semi-solid heavy oil. Its color ranges all the way from a light golden yellow to a deep dark black (Fig. 6.1). Oil from different geographical locations has its own very unique properties like viscosity, volatility, color, and density. Crude oil is a complex liquid mixture made up of a vast number of hydrocarbon compounds, and small amounts of organic compounds containing sulfur, oxygen, nitrogen and metals such as vanadium, nickel, iron, and copper. The commercial price of a crude oil is impacted by many of its properties, one of which is called *gravity*, generally expressed as **API gravity** = (141.5)/(its specific gravity) – 131.5, where specific gravity of a crude oil is (its density)/(density of water) at 60 $°F$. The API gravity of crude oils ranges from <8.5 for very heavy crudes, 8.5 and 29 for heavy crudes, 29–38 for medium crudes, and >38 for light crudes. There are over 160 different crude oil types traded in the world market these days. The three primary ones among them are:

WTI (West Texas Intermediate) is a premium quality crude oil, greatly valued because more and better gasoline can be made from a single barrel of this than most other types available on the market. It is a light, sweet (with only 0.24% sulfur by weight) crude oil. Its price in the market is often $ 5–7 higher per barrel than OPEC basket price.

Brent Blend is a combination of different crude oils from 15 fields in the North Sea; it is light and sweet (0.37% sulfur), excellent for making gasoline and the middle distillates, typically priced $ 4 per barrel higher than OPEC basket price.

OPEC Basket is a group of seven different crude oils from Algeria, Saudi Arabia, Indonesia, Nigeria, Dubai, Venezuela, and the Mexican Isthmus, not as sweet or as light as the above two groups.

Refineries can accommodate a different crude slate in one of two ways. If the properties of the crude vary only slightly, then process cut points, charge rate, and operating set points of existing units can accommodate the change through modifications in operating conditions. However, when there are more than minor changes in crude properties, refineries require finding a blend of several available crude oils that comes quite close in properties as the current slate.

We discuss procedures used in industry for crude oils blending in the next chapter.

The other is mixing different output products from the CDU and other treatment processes in the crude oil refinery, and other additives like octane rating enhancers, metal deactivators, antioxidants, anti-knock agents, rust-inhibitors, and detergents and other components to produce finished products with specified properties and other desired characteristics to sell in the markets.

Products can be blended in-line by injecting proportionate amounts of each component in the main stream where turbulence promotes thorough mixing, or blended by batches in tanks and vessels.

In product blending for example, gasoline is produced by blending a number of products including alkylate (output from alkylation units), reformate (upgraded naphtha resulting from catalytic or thermal reforming), fluid catalytic cracking gasoline, and oxygenated additives such as Methyl Tertiary Butyl Ether (MTBE) to increase the octane rating.

In this chapter we will discuss the product blending problems, and the mathematical models used for finding optimum blending recipes for each of them; and how optimum solutions are computed for them, and implemented in practice.

Final quality of the outputs from blending are always checked by lab tests before dispatch to the markets. Here are brief descriptions of these properties from [1].

Pour point (PP) of an oil product is defined as the lowest temperature at which a sample of this oil product will flow. It indicates how difficult it is to pump this oil product in cold weather. Lower pour point means that the paraffin content in it is low.

Viscosity (VIS) of an oil product determines the resistence of this oil product to flow. More viscous oils create a greater pressure drop when they flow in pipes.

Refractory index of an oil product is the ratio (velocity of light in a vacuum)/(velocity of light in the oil product), used as a characterization parameter for the product's composition.

Freezing point of an oil product is the temperature at which the hydrocarbon mixture solidifies at atmospheric pressure, an important property for kerosene and jet fuels due to the very low temperatures encountered at high altitudes by jet planes.

Aniline point (AP) of an oil product is the lowest temperature at which an equal volume mixture of the oil product and aniline are miscible. Since aniline is an aromatic compound, a petroleum product with high aromatic content will be miscible at ambient conditions. Oil products with more paraffins will have higher aniline points. This is an important specification for diesel fuels.

Flash point of a liquid hydrocarbon mixture is the lowest temperature at which sufficient vapors are produced above the liquid to cause a spontaneous ignition in the presence of a spark. An important specification for gasoline and naphtha, related to safety in storage and transport in high temperature environments, and an indicator of fire and explosion potential of a liquid fuel; a lower flash point fuel is a higher fire hazard.

Octane number (ON) is a measure of knocking tendency of gasoline fuels in spark ignition engines, which is measured under two environments. Motor Octane Number (MON) indicates engine performance at highway conditions with high speeds, Research Octane Number (RON) is indicative of performance at low speed city driving. The Posted Octane Number (PON) is the average of MON and RON.

Cetane number (CN) of a diesel fuel is the percentage of pure cetane (n-hexadecane) in a blend of cetane and alpha methyl naphthalene which matches the ignition quality of the diesel fuel sample.

Smoke point is the maximum height in mm; of a smokeless flame of fuel, it measures the burning qualities of kerosene and jet fuel.

Reid vapor pressure (RVP) of a liquid oil product is the vapor pressure determined in a volume of air four times the liquid volume at $37.8°C$ $(= 100°F)$. It measures the vapor-lock tendency of a motor gasoline in which excessive vapors in the fuel line causing interruption of the supply of liquid fuel to the engine. It indicates the evaporation and explosion hazards of the fuel.

Blending of Output Products from the CDU and Other Treatment Processes into Finished Products for the Marketplace

Output products from treatment processes are blended into finished products like gasoline, kerosenes, and diesel oil for the marketplace. Here we discuss the mathematical models used to determine the optimum blending recipes for making these finished products to meet the specifications on various properties on them.

Gasolines are checked for octane rating, Reid Vapor Pressure (RVP), and volatility. Kerosenes are tested for flash point, volatility. Gas oils (middle distillate petroleum fractions with a boiling point in the range 350–750°F including diesel fuel, kerosene, heating oil, light fuel oil, etc.) are tested for pour point, and viscosity.

Properties (such as octane number, specific gravity, RVP, viscosity, flash point, pour point, cloud point, and aniline point) of a blended product can be determined as functions of the corresponding property of the components in the blend, and the quantities of the components in the blend. Consider a blend with the following data corresponding to volumes in barrels of the components blended:

n = number of components in the blend.

$i = 1$ to n is the index representing the components in the blend.

x_i: $i = 1$ to n is the proportion (volume fraction) of compound i in the blend. So $\sum_{i=1}^{n} x_i = 1$.

p_i: $i = 1$ to n is the value of a property for component i in the blend.

If the property is linear (i.e., additive) under blending, then the value of the property for the blend will be $\sum_{i=1}^{n} x_i p_i$. Additive properties include specific gravity, boiling point, and sulfur content. However, properties like viscosity, flash temperature, pour point, aniline point, RVP, cloud point, and octane number are not additive. Petroleum industry has developed techniques for getting estimates of the property values of nonadditive properties of petroleum component blends. One such empirical method developed by **Chevron oil trading co.** is based on "**blending indices**", see [1, 2]. These blending indices are specific to the property, for each component in the blend, whether the blending calculations are being performed in terms of proportions by volume or weight; and for each nonadditive property there are separate blending indices. We will now discuss these blending indices and how to use them for estimating each property of the blend, see [1]. We will show the procedures when blending calculations are performed in terms of proportions by volume. Corresponding procedures exist if the preference is to do these blending calculations using proportions by weight. We will discuss these blending indices (BI) for each property.

Reid Vapor Pressure (RVP) of the Blend: RVP of an oil product is the vapor pressure in psi (it is a unit of pressure expressed in pounds of force per square inch) at $100°F$ determined in a volume of air four times the liquid volume; it is not an additive property. The blending index used for RVP, $BI_{RVP,i} = (RVP_i)^{1.25}$ is the blending index for component i in the blend in psi, where RVP_i is the RVP in psi for component i in the blend. And the estimate $BI_{RVP,Blend}$ of the blend is $\sum_{i=1}^{n} x_i BI_{RVP,i}$, and hence the estimate of the RVP of the blend is $(BI_{RVP,Blend})^{1/1.25}$.

Example 1 Consider a blend consisting of components: LSR gasoline, HSR gasoline, Reformate, and FCC gasoline with the following data. Quantities are measured in BPD (barrels per day).

Quantities and RVP values of components blended

Component $i =$	Quantity (BPD)	x_i	RVP_i in psi	$BI_{RVP,i}$
1. LSR gasoline	5000	0.227	11.1	20.26
2. HSR gasoline	4000	0.182	1.0	1.0
3. Reformate 94 RON	6000	0.273	2.8	3.62
4. FCC gasoline	7000	0.318	13.9	26.84

So here $n = 4$, $p_i = RVP_i$ and $BI_{RVP,i}$ in psi are given in the table. Hence $\sum_{i=1}^{4} x_i BI_{RVP,i} = 14.305$. Consequently the RVP of the blend is $(14.305)^{1/1.25} = 8.4$ psi.

Example 2 Additives such as i-butane, n-butane can be added to the blend in Example 1 to increase its RVP. Consider adding n-butane with RVP of 52 psi to the blend prepared in Example 1 to get a mixture with RVP = 10 psi. How much n-butane should be added? Let λ denote the BPD of n-butane added to the blend prepared in Example 1 to achieve this goal. Then we have the following data for the new mixture.

Quantities and RVP values of components blended

Component $i =$	Quantity (BPD)	x_i	RVP_i in psi	$BI_{RVP,i}$
1. LSR gasoline	5000	$5000/(22000 + \lambda)$	11.1	20.26
2. HSR gasoline	4000	$4000/(22000 + \lambda)$	1.0	1.0
3. Reformate 94 RON	6000	$6000/(22000 + \lambda)$	2.8	3.62
4. FCC gasoline	7000	$7000/(22000 + \lambda)$	13.9	26.84
5. n-butane	λ	$\lambda/(22000 + \lambda)$	52	139.64

So, the requirement that the RVP of the mixture should be 10 leads to the equation $(10)^{1.25} = (5000 \times 20.26 + 4000 \times 1.0 + 6000 \times 3.62 + 7000 \times 26.84 + \lambda \times 139.64)/(22000 + \lambda)$

i.e., $10^{1.25}(22000 + \lambda) = 314912 + 139.64 \times \lambda$

This leads to $\lambda = 625.7$ BPD which is the amount of n-butane to add to the blend in Example 1 to bring the RVP of the mixture to 10 psi.

Pour Point (PP) of the Blend: The pour point is the lowest temperature at which an oil product can be stored and it is still capable of flowing or pouring without stirring, when it is cooled. It is not an additive property, and so PP blending indices are used which blend linearly when quantities of components in the blend are measured on a volume basis. Here the temperature of an object can be measured in three different scales: **Celcius, Kelvin, Rankin**. Let $[\,^{oC}]$, $[\,^{oK}]$, $[\,^{oR}]$ denote the temperature of the object in these three scales in that order. Here are the relationships between these measurements: $[\,^{oK}] = [\,^{oC}] + 273.15$; $[\,^{oR}] = ([\,^{oC}] + 273.15)(9/5)$; $[\,^{oC}] = ([\,^{oR}] - 491.67)(5/9)$.

Let PP_i be the pour point in $[\,^{oC}]$ of the ith component in the blend. First convert the PP temperatures of all the components and measure them in ^{oR}. Then the blending index $BI_{PP,i}$ of component i in the blend in ^{oR} is

$BI_{PP,i} = 3262000(PP_i/1000)^{12.5}$

The estimate of the blending index of the blend, $BI_{PP,Blend} = \sum_{i=1}^{n} x_i BI_{PP,i}$ and hence the estimate of the PP_{Blend} is $1000[(BI_{PP,Blend})/3262000]^{1/12.5}$ in ^{oR}. Convert it back into ^{oC} yielding the estimate of the PP of the blend in ^{oC}.

Example 3 Consider a blend consisting of four components with data given in the following table.

Quantities and PP values of components blended

Component i =	Quantity (BPD)	x_i	PP_i in ^{OC}	PP in ^{OR}	$BI_{PP,i}$
1. Catalytic cracked gas oil	2000	0.1818	−15	464.67	227.3
2. Straight run gas oil	3000	0.2727	−3	486.27	461.0
3. Light vacuum gas oil	5000	0.4545	42	567.27	2748
4. Heavy vacuum gas oil	1000	0.0909	45	572.67	3093

So, $\sum_{i=1}^{4} x_i BI_{PP,i} = 1697.2$. So, $PP_{Blend} = [(1697.2)/(3262000)]^{1/12.5} \times 1000 = 546^{0R}$. Converting back to ^{oC} we get $PP_{Blend} = 30^{oC} = 86^{oF}$.

Note: An alternate formula for $BI_{PP,i}$ is also being used in the industry in recent times; it is $BT_{PP,i} = (PP_i)^{1/(0.08)}$. As an exercise compute the PP_{Blend} in this example using this formula.

Cloud Point (CP) of the Blend: Cloud point (CP) of an oil product is the lowest temperature at which the oil becomes cloudy and the first particles of wax crystals are observed as the oil is cooled gradually under normal conditions. It is not an additive property, so CP blending indices are used, which blend linearly when quantities of components included in the blend are measured on a volume basis, to estimate the CP of the blend.

We are given the CP of component i included in the blend, for $i = 1$ to n in ^{oC}. If $[\,^{o}K]$, $[\,^{oC}]$ denote temperature in the Kelvin, Centigrade scales, then $[\,^{oK}] = [\,^{oC}] +$

273.15, [^{oC}] = [^{oK}]—273.15. For $i = 1$ to n, convert the CP of component i included in the blend into the Kelvin scale, and let CP_i be the cloud point of component i in the blend in ^{oK}.

Cloud point is not an additive property, so we use cloud point blending indices, $BI_{CP,i}$, which blend linearly when quantities of components included in the blend are given on a volume basis. The $BI_{CP,i} = (CP_i)^{1/0.05}$. So $BI_{CP,Blend} = \sum_{i=1}^{n} x_i BI_{CP,i}$.

From this we can calculate $CP_{Blend} = (BI_{CP,i})^{0.05}$, in ^{oK}, and we convert it into ^{oC}.

Example 4 Consider a blend with three components with the following data.

Quantities and CP values of components blended

Component i =	Quantity (BPD)	x_i	CP_i in ^{oF}	CP_i in ^{oK}	$BI_{CP,i}$
1. A	1000	0.167	−5	252.4	1.101×10^{48}
2.B	2000	0.333	5	258.0	1.701×10^{48}
3.C	3000	0.5	10	260.8	2.119×10^{48}

So $BI_{CP,Blend} = \sum_{i=1}^{3} x_i BI_{CP,i} = 1.8098 \times 10^{48}$. So, $CP_{Blend} = (1.8098 \times 10^{48})^{0.05}$ in ^{oK} = $258.756^{oK} = -14.25^{oC} = 6.35^{oF}$.

Aniline Point (AP) of the Blend: AP of an oil is the minimum temperature at which equal volumes of that oil and pure aniline are completely miscible, and indicates the degree of aromaticity of the oil. It is not an additive property, and so blending indices which blend linearly when proportions of components in the blend are given on a volume basis.

Let AP_i be the AP of component i in the blend in ^{oC}. Then the AP of the ith component in the blend, $BI_{AP,i} = 1.124[exp(0.00657AP_i)]$. Then $BI_{AP,Blend}$, the blending index of the blend = $\sum_{i=1}^{n} x_i BI_{AP,i}$, where x_i is the proportion by volume of the ith component in the blend. And $AP_{Blend} = log_n(BI_{AP,BLend}/1.124)/0.00657$ in ^{oC}.

Example 5 Consider a blend of three components with the following data:

Quantities and AP values of components blended

Component i =	Quantity (BPD)	x_i	AP_i in ^{oC}	$BI_{AP,i}$
1. Light Diesel	4000	0.4	71.0	1.792
2. Kerosene	3000	0.3	60.7	1.675
3. Light cycle gas oil	3000	0.3	36.8	1.432

Then $BI_{AP,Blend} = \sum_{i=1}^{3} x_i BI_{AP,i} = 1.6468$. So, $AP_{Blend} = log_n(\sum_{i=1}^{3} x_i BI_{AP,i}/1.124)/0.00657 = 58.17^{oC}$.

Smoke Point (SP) of the Blend: SP of an oil is the maximum flame height in mm at which the oil burns without smoking when tested at specified standard conditions.

Consider the blend in terms of volume fractions discussed earlier under aniline point of the blend in Example 5. The SP of that blend in mm is SP_{Blend} determined using the following formula:

$SP_{Blend} = -255.26 + 2.04AP_{Blend} - 240.8log_n(SG_{Blend}) + 7727(SG_{Blend}/AP_{Blend})$

where the AP_{Blend}, SG_{Blend} are the aniline point and specific gravity, respectively, of that blend. The specific gravity is an additive property, hence the specific gravity of a given blend with component fractions by volume in the blend x_i, and specific gravities of components as SG_i, is $SG_{Blend} = \sum_{i=1}^{n} x_i SG_i$.

Example 6 The specific gravities of components in the blend discussed in Example 5, SG_i, are 0.75, 0.8, 0.8, respectively. So the specific gravity of that blend is $SG_{Blend} = 0.4(0.75) + 0.3(0.8) + 0.3(0.85) = 0.805$, and the smoke point of that blend $SP_{Blend} = -255.26 + 2.04(58.17) - 240.8log_n(0.805) + 7727(0.805/58.17) = 22.33$ mm.

Viscosity of the Blend: Viscosity is not an additive property, so viscosity blending indices $BI_{Vis,i}$ are used for determining the viscosity of a blend, V_{Blend}. Viscosity is one of an oil's most important physical properties; it is often the first parameters measured by most oil analysis labs because of its importance to oil condition and lubrication. It is measured in units denoted by "cSt".

Let v_i = viscosity of component i in the blend, in cSt. Then the blending index of component i in the blend is $BI_{Vis,i} = (log_{10}v_i)/(3 + log_{10}v_i)$. Then

$BI_{Vis,Blend} = \sum_{i=1}^{n} x_i BI_{Vis,i}$

and the viscosity of the blend is $v_{Blend} = 10^{3(BI_{Vis,Blend})/(1-BI_{Vis,Blend})}$ in cSt.

Example 7 Consider a blend of three components with the following data:

Quantities and Viscosity values of components blended

Component i =	Quantity (BPD)	x_i	V_i in cSt	$BI_{Vis,i}$
1 Fraction 1	2000	0.222	75	0.385
2. Fraction 2	3000	0.333	100	0.400
3. Fraction 3	4000	0.445	200	0.434

Here blending index $BI_{Blend} = \sum_{i=1}^{3} x_i BI_{Vis,i} = 0.4117$. So, the viscosity of the blend is $V_{Blend} = 10^{3(0.4117)/(1-0.4117)} = 125.73$ cSt.

Flash Point (FP) of the Blend: The flash point of an oil is the lowest temperature at which the vapors coming from the oil ignite. It indicates the maximum temperature at which the oil can be stored safely without causing a serious hazard. If the flash point of a petroleum product is considered to be high, it can be blended with other fractions to lower it. FP is not an additive property, and so flash point blending indices are used which blend linearly when volume fractions of components in the blend are used. Let x_i, FP_i be the volume fraction, flash point temperature of the ith component in the blend in ^{o}K. Then the blending index $BI_{FP,i} = FP_i^{1/(-0.06)}$.

Example 8 Consider a blend of three components A, B, C with the following data:

Quantities and Flash point values of components blended

Component i =	Quantity (BPD)	x_i	FP_i in ^{oK}	$BI_{FP,i}$
1 A	2500	0.222	120	2.22×10^{-35}
2 B	3750	0.333	100	4.64×10^{-34}
3 C	5000	0.445	150	5.39×10^{-37}

The resulting BI_{FP} of the blend is 1.60×10^{-34}. Numerically inverting the conversion formula $BI_{FP} = FP^{1/(-0.06)}$ yields an FP for the blend of 106.6^{oK}.

Octane Number (ON) of a Gasoline Blend: The octane number (ON) is an important characteristic of fuels used for spark ignition engines. ON is a measure of the fuel's tendency to "knock" in an engine test. Post octane number (PON) of a fuel, also known as the road octane number, is the average of the fuel's RON and MON.

There are several additives such as oxygenated ethers or alcohols that can be added to gasoline to enhance its octane number. Some like tetra-ethyl lead used in the past have now been phased out, but other oxygenates like methanol, ethanol, Methyl tertiary butyl ether (MTBE), Ethyl tertiary butyl ether (ETBE), Tertiary butyl alcohol (TBA), Tertiary amyl methyl ether (TAME) are still being used in practice for this purpose.

The octane number discussed below, ON, may be either RON, MON, or PON; the procedure described below for estimating the octane number of a blend with the given data applies to all of them.

Let ON_i, x_i be respectively the octane number, the fraction by volume of the ith component in the blend. If the octane number is an additive property under blending in terms of volume fraction, the octane number of the blend ON_{Blend} will be = $\sum_{i=1}^{n} x_i ON_i$. But in practice it has been observed that this estimate needs minor corrections to get the true value. Blending indices for components in the blend are used to get a closer estimate for the octane number, ON_{Blend}, of the blend.

The blending index of a component in the blend, is a function BI_{ON} of its octane number ON. This blending index function BI_{ON} is a piecewise nonlinear function of ON. The range of values of ON is divided into three intervals and in each interval BI_{ON} is defined as given below:

$11 \leq ON \leq 76, \quad BI_{ON} = f_1(ON) = 36.01 + 38.33(ON/100) - 99.8 (ON/100)^2 + 341.3(ON/100)^3 - 507.02(ON/100)^4 + 268.64(ON/100)^5$

$76 \leq ON \leq 103, \quad BI_{ON} = f_2(ON) = -299.5 + 1272(ON/100) - 1552.9 (ON/100)^2 + 651(ON/100)^3$

$103 \leq ON \leq 106, \quad BI_{ON} = f_3(ON) = 2206.3 - 4313.64(ON/100) + 2178.57(ON/100)^2$

Let R_1, R_2, R_3 be the interval of values of BI_{ON} corresponding to the three intervals of ON above. Actually, the function $f_3(ON)$ is commonly used to determine the value of BI_{ON} even when ON is greater than 106.

Then the blending index for the blend, $BI_{ON,Blend} = \sum_{i=1}^{n} x_i BI_{ON_i}$. To get the estimate ON_{Blend} of the octane number of the blend, we find which of the three intervals R_1, R_2, R_3 contains the value of $BI_{ON,Blend}$ and if it is R_j, then ON_{Blend} is the value of ON that satisfies $f_j(ON) = BI_{ON,Blend}$.

Example 9 Consider a blend of three components A, B, C with the following data:

Quantities and RON values of components blended

Component i	Quantity (BPD)	x_i	RON_i	$BI_{RON,i}$
1 Alkylate $c_4^=$	6000	0.4	97.3	67.66
2 Coker gasoline	4000	0.267	67.2	53.65
3 FCC gasoline	5000	0.333	83.2	58.78

If the octane number is an additive property under blending in terms of volume fraction, the octane number of the blend ON_{Blend} will be $= \sum_{i=1}^{n} x_i ON_i = 84.57$

Using the blending index method, $BI_{RON,Blend} = 0.4(67.66) + 0.267(53.65) + 0.333(58.78) = 60.98$. 60.98 is contained in the interval R_2 in this example. So, RON_{Blend} is the solution of $f_2(RON) = 60.98$, which leads to the estimate $RON_{Blend} = 88.09$.

Example 10 Consider the blending problem discussed in Example 9. Suppose it is required to produce a gasoline with a RON of 95 by adding the oxygenate MTBE with RON of 115 to the blend in Example 9. Find out how much MTBE should be added to that blend.

Let V_{MTBE} denote the BPD of MTBE added to the blend discussed in Example 9 to meet this requirement; and let BLEND denote this new blend. So V_{BLEND}, the total volume of this BLEND, is $15000 + V_{MTBE}$ BPD.

We want the RON value of BLEND to be 95; i.e., using the blending index method we want $BI_{RON,BLEND}$ to be $f_2(95) = 65.56$. Since MTBE has a RON of 115, we get $BI_{RON,MTBE} = f_3(115) = 126.77$. So, the requirement $BI_{RON,BLEND} = 65.56$ becomes the equation

$67.76(6000/V_{BLEND}) + 53.65(4000/V_{BLEND}) + 58.78(5000/V_{BLEND}) + 126.77(V_{MTBE}/V_{BLEND}) = 65.56$

or $65.56(15000 + V_{MTBE}) = 6000(67.66) + 4000(53.65) + 5000(58.78) + 126.77 V_{MTBE}$

leading to $V_{MTBE} = 1126$ BPD.

Method to Find an Optimum Blending Recipe Using Blending Indices

Here we will consider the problem of finding an optimum blending recipe to blend components from a given set of petroleum components, to prepare a blend satisfying specified constraints on various properties like octane number, etc., using blending indices. A solution called a *blend* is a vector $x = (x_i)$ satisfying $\sum_{i=1}^{n} x_i = 1$, where x_i is the proportion by volume of component i in the blend; and these are the decision variables.

Example 11 Consider seven components with the following data (partially adapted from Example E9.14 of reference [1]):

RVP and RON values of various components available to be blended

Component i	RVP_i (psi)	RON_i	$Cost_i$ ($ per barrel)
1. n-Butane	51.6	93	38
2. Isomer gasoline	13.5	83	42
3. Reformer gasonline	2.2	98.5	45
4. FCC gasoline	4.4	92.3	38
5. Hydrocracker gasoline	12.9	82.8	33
6. Alkylate	4.6	97.3	44
7. Polymer	8.7	96.9	41

Suppose that the goal is to create a minimum cost blend with an RVP of 10.2 psi and a RON of 89. We first convert the RVP and RON values to their corresponding BI values:

Conversions to BIs.

i	BI_{RVP_i} (psi)	BI_{RON_i}
1	$(51.6)^{1.25} = 138.3$	$f_2(93) = 64.0$
2	$(13.5)^{1.25} = 25.9$	$f_2(83) = 58.7$
3	$(2.2)^{1.25} = 2.7$	$f_2(98.5) = 68.9$
4	$(4.4)^{1.25} = 6.4$	$f_2(92.3) = 63.5$
5	$(12.9)^{1.25} = 24.4$	$f_2(82.8) = 58.6$
6	$(4.6)^{1.25} = 6.7$	$f_2(97.3) = 67.7$
7	$(8.7)^{1.25} = 14.9$	$f_2(96.9) = 67.3$

We then formulate the following LP:

$$\min_x \sum_{i=1}^{7} x_i \text{Cost}_i$$

$$\text{s.t.} \sum_{i=1}^{7} x_i \text{BI}_{\text{RVP}_i} = \text{BI corresponding to an RVP of 10.2 psi} = (10.2)^{1.25}$$

$$\sum_{i=1}^{7} x_i \text{BI}_{\text{RON}_i} = \text{BI corresponding to a RON value of 89} = f_2(89)$$

$$x_i \geq 0, \ i = 1, 2, \ldots, 7$$

$$\sum_{i=1}^{7} x_i = 1$$

The resulting optimal solution is a blend of 24.3% FCC gasoline, 56.5% Hydrocracker gasoline, and 19.1% Polymer, at a total cost of $35.75 per barrel.

Exercise 1 Consider four components with the following data:

Barrels available (maximum), RVP, and Viscosity values of various components

Component i	Barrels (max)	RVP_i (psi)	$Viscosity_i$ (cSt)	$Cost_i$ ($ per barrel)
1. LSR gasoline	3000	21.1	0.2886	30
2. Hydrocracker gas	2000	15.5	0.4140	33
3. FCC gasoline	6025	4.8	0.8607	38
4. n-Butane	640	51.6	0.3500	38

(a) Use blending indices to model the problem of finding a minimum cost blend of 5000 total barrels of a mixture with an RVP of 12 psi and a viscosity of 0.525 cSt.

(b) Use blending indices to model the problem of finding a minimum cost blend of 7000 total barrels of a mixture with an RVP of 12 psi and a viscosity of 0.525 cSt.

In the industry, some products (such as gasoline and diesel) are blended by volume. Some products (such as fuel oil) are blended by weight. There is seldom one "best" formula for blending indices, and different oil companies use different formulae. One such company has some free software: Google "Haverly Property Calculator" that can be downloaded. Unfortunately, companies are very secretive about the formulae that they use for blending indices.

Reference

1. "Fundamentals of Petroleum Refining", Mohamed A. Fahim, Taher A. Alshahhaf, and Amal Elkilani, Elsevier, Netherlands, 2010.
2. "Optimizing Chevron's Refineries", Ted Kutz, Mark Davis, Robert Creek, Nick Kenaston, Craig Stenstrom, Margery Connor, INFORMS Journal on Applied Analytics, Vol. 44, No. 1 , January-February 2014, Pages 39-54, Winner of 2014 Franz Edelman Award of INFORMS.

Chapter 7
Crude Oils Blending Problem in Oil Refineries to Prepare Feedstock into the Crude Distillation Unit

Katta G. Murty

Abstract Some of the crude oil sources reaching the end of their lifecycles, and International Politics, play major roles in the supply of crude oils to an oil refinery. In this chapter, we discuss the procedures used by refineries to source what is available at best prices, and blend them into their feedstock optimally.

With some of the old crude oil sources getting depleted and reaching the end of their life cycles, nowadays many oil refineries are forced to get their crude oil supply from multiple sources. Also, **International Politics can play** a major role in the supply of crude oils to a refinery. Under these circumstances, refineries are forced to source what is available at best prices, and figure out a blend of those that come close to the refinery's process capabilities. Crude oils from different sources have different properties like color, density, viscosity, composition, pour point, sulfur content, acidity, and sediment content.

Typically crude oil is classified as light, medium, and heavy according to its API gravity being ≥ 31, between $22 - 31$, and ≤ 22, respectively (these ranges for different classification labels are not unique, they vary slightly depending on who you talk to). The contents of a crude oil have a large impact on its price. Light crudes normally fetch a higher price than heavy crudes. Some refineries may favor light crudes given the higher percentage of components that without further refining, supplies material for the higher profit, higher demand gasoline pool. But this favoritism has a limit; if price differential is high enough, the refinery may decide to build secondary process units to further refine heavy materials from heavy crude despite the cost, because the savings in crude price more than offset the increased processing costs of additional units.

For an established refinery, typically, they are operated so as to maximize their net profit/day. This can be achieved by blending high cost light crude oils with low cost heavy crude oils to produce a blend with properties satisfying the operating range

K. G. Murty (✉)
Department of IOE, University of Michigan, Ann Arbor, MI48109-2117, USA
e-mail: murty@umich.edu

K. G. Murty (ed.), *Models for Optimum Decision Making*, International Series in Operations Research & Management Science 286, https://doi.org/10.1007/978-3-030-40212-9_7

of the refinery. Blending of crude oils from different sources helps take advantage of refinery process flexibility in feedstock properties, by enabling to mix low cost crudes which as individual feeds do not fall into the operating range of the refinery; into a blend meeting the requirements. This blend of available crude oils should have physical and chemical properties suitable for smooth operation of distillation unit at minimum cost.

Crude oil contains the inorganic acid substance H_2S (Hydrogen Sulfide), and organic acids such as naphthenic acid. These make the crude oil acidic. These acids react with Potassium Hydroxide (KOH) in a chemical reaction. To measure the **Total Acid Number (TAN)** of a crude oil, they measure how many mg (milligrams) of KOH are needed to neutralize the acid in one liter (l) of crude oil; that is why the TAN of a crude oil is also referred to as its **TAN mg KOH/l**.

According to [1], heavier crude oils are more expensive to distill than light crude oils, as they require more energy in heating to their boiling point for distillation, and they will have a higher bottom residue for further distillation in vacuum distillation. High sulfur content in crude oil poisons the catalyst in catalytic processes in refining, causing refineries to purchase larger amounts of catalysts; also high sulfur content in crude oil increases the rate of corrosion of refinery units and pipelines. And higher Total Acid Number (TAN) in crude oil processed, results in enhanced corrosion of the CDU increasing the cost of maintenance.

For reasons similar to these, at most refineries, they have target specification ranges for some properties of their feedstock into their CDU. The most important of these quality characteristics are API gravity (measured using density), sulfur content as a % by weight, and Total Acid Number (TAN); at most refineries they have specifications for the ranges of these properties in their feedstock. Some refineries may also have specifications on other characteristics like pour point for their feedstock. The refinery specifies the ranges of these properties that its units can easily adopt to by slightly adjusting process temperature requirements such as preheating, and different distillation temperatures. Estimating the correct values of these properties for a given blend of available crude oils requires complex nonlinear functions. For nonlinear, Mixed-integer Linear Programming models for crude oils blending problem, see references [2, 3].

But it has been found that linearity assumption leads to approximate estimates that are satisfactory in practice [1, 4, 5]. So in industry, unlike in product blending, estimating the properties of a blend in terms of the same properties of crude oils going into the mixture is based on the linearity assumption.

So the general problem of blending available crude oils into the feedstock for the CDU of a refinery is a linear program (LP) of the following form: Let

k = number of different properties of the feedstock on which there are specifications set by the refinery,

ℓ = number of different crudes available to be blended into the feedstock,

i, j = indices we use to indicate different properties to be satisfied, and crude oils available,

u_i, v_i = lower and upperbounds specified for property i, $i = 1$ to k,

q_{ij} = value of property i for crude oil j available, for $i = 1$ to k, $j = 1$ to ℓ,

c_j = price of available crude oil j in US \$/barrel,

w = total required number of barrels of feedstock needed during the period,

a_j, b_j = lower and upperbounds on the number of barrels of crude oil j available during this period, to be used in the feedstock, for $j = 1$ to ℓ,

x_j = fraction of available crude oil j, in the blend into feedstock, and

$x = (x_j : j = 1 \text{ to } \ell)$ = vector of blending fractions of the available crude oils in the mix used.

Then the model for finding the optimum x is the following LP:

Minimize $\sum_{j=1}^{\ell} c_j w x_j$

subject to $u_i \leq \sum_{j=1}^{\ell} q_{ij} x_j \leq v_i, \quad i = 1 \text{ to } k$

$a_j \leq w x_j \leq b_j, \quad j = 1 \text{ to } \ell$

$\sum_{j=1}^{\ell} x_j = 1$

$x_j \geq 0, \quad j = 1 \text{ to } \ell$

The optimum solution to this model, $\bar{x}$, is the vector of optimum blending fractions of the available crude oils to prepare the feedstock.

Data on Some Available Crude Oils

The following table taken from reference [1] gives data on properties of 30 crude oils available in the market. The estimated market price in US \$/barrel of a crude oil is denoted by "M (Brent price)" where the factor "M" is called the **price factor** in the following table. In the scale used in this table, light crudes, medium crudes, and heavy crudes have API gravity $\geq$31, between 22–31, and $\leq$10, respectively (Table 7.1).

Let M_j be the price factor for available crude oil j, for $j = 1$ to ℓ. Then the objective function for minimization in the above LP is $(\sum_{j=1}^{\ell} M_j x_j)$ (w (Brent price)). So, its optimum solution is the same as that for minimizing $(\sum_{j=1}^{\ell} M_j x_j)$ subject to the same constraints.

Example: An oil refinery sources their crude oil from five sources; they are Saharan Blend, Iranian Heavy, Bow River Hardisty, Syrian Heavy, and Azeri Light. The specifications for the feedstock into their CDU, for properties API gravity, Sulfur, TAN are 30.2 ± 2.0, 1.5 ± 0.5, and 0.4 ± 0.2, respectively. Find the optimal blending mix of these five crude oils to form the feedstock into their CDU, minimizing z = the cost of the inputs from the five sources, to prepare w = 10 million barrels of feedstock. Each of the components going into the mixture are available upto 9 million barrels. And the Brent price at that time is 63 US dollars/barrel. As discussed above, let x_j = fraction of input j available, used in feedstock, for $j = 1$ to 5.

The optimum solution of the linear programming model for this example is the same as the optimum solution of

Minimize $z = (1.0452x_1 + 0.9282x_2 + 0.8572x_3 + 0.8374x_4 + 0.9949x_5)$

Subject to $28.2 \leq 45.3x_1 + 30.1x_2 + 20.3x_3 + 23.1x_4 + 34.8x_5 \leq 32.2$

$1.0 \leq 0.12x_1 + 1.78x_2 + 2.96x_3 + 4.2x_4 + 0.15x_5 \leq 2.0$

$0.2 \leq 0.37x_1 + 0.13x_2 + 0.69x_3 + 0.28x_4 + 0.2x_5 \leq 0.6$

$x_1 + x_2 + x_3 + x_4 + x_5 = 1$

$x_j \geq 0$ for $j = 1$ to 5.

Table 7.1 Data on some available crudes from [1]

Crude	API gravity	Sulfur % by weight	Tan mg KOH/l	*M*
1. Brent	37.5	0.4	0.041	1.0
2. Kumkol	42.5	0.07	0.1	1.0333
3. Zarzatine	42.6	0.08	0.23	1.0334
4. Brass river	37.4	0.11	0.06	1.0081
5. Saharan blend	45.3	0.12	0.37	1.0452
6. Qualboe	36.0	0.13	0.26	1.0007
7. Azeri light	34.8	0.15	0.2	0.9949
8. Bonny light	35.3	0.15	0.5	0.9972
9. Cabinda	37	0.17	0.03	1.0045
10. Escravos	33.5	0.17	0.104	0.9884
11. Ekofisk	38.5	0.19	0.56	1.0109
12. Escalante	24.1	0.19	0.26	0.9460
13. Oseberg	37.8	0.27	0.4	1.0051
14. Forcados	30.4	0.28	0.1	0.9713
15. WTI	40.8	0.34	0.1	1.0171
16. Esider	36.7	0.37	0.01	0.9972
17. Siberian light	37.8	0.42	0.652	1.008
18. CPC blend	45.3	0.56	0.06	1.0317
19. Syrian light	38.0	0.68	0.05	0.9942
20. Forties	38.7	0.79	0.093	0.9942
21. Flotta	36.2	0.98	0.15	0.9773
22. Ural	31.3	1.25	0.08	0.9480
23. Iranian light	33.1	1.33	0.09	0.9536
24. Suez blend	31.3	1.41	0.06	0.9435
25. Oriente	24.0	1.59	0.04	0.9076
26. US poseidon	29.7	1.65	0.41	0.9301
27. Iranian heavy	30.1	1.78	0.13	0.9282
28. Kirkuk	34.3	2.28	0.09	0.9326
29. Bow river hardisty	20.3	2.96	0.69	0.8572
30. Syrian heavy	23.1	4.2	0.28	0.8374

We solved this model using the Excel Solver. The optimum solution of this model is $(x_1, x_2, x_3, x_4, x_5) = (0, 0.79392, 0.197833, 0, 0.008247)$ with the optimum objective value = 0.914704. So, the minimum cost of crude inputs in the blend for the feedstock of 10 million barrels is = 10 (63)(0.914704) = 576.261 million US dollars, and the optimum blend consists of 79.392% of Iranian Heavy, 19.7833% of Bow River Hardisty and 0.8247% of Azeri Light crude oils.

Exercise 1 (Geoffrey Gill): In the optimum solution of this example, the Saharan Blend, whose market price is US $ 1.0452 (Brent price), is not used. Assuming that all the data other than the market price of this Saharan Blend remains the same, how much should the market price of this Saharan Blend decrease before it appears in an optimum solution of this example?

Exercise 2: An oil refinery sources their crude oil from 5 sources, they are Flotta, Ural, Ekofisk, Bow River Hardisty, Siberian Light, and Suez Blend. The specifications for the feedstock into their CDU, for properties API gravity, Sulfur, and TAN are 34 ± 4.0, 1.6 ± 0.5, and 0.5 ± 0.3, respectively. Each of the components going into the mixture are available upto 10 million barrels. And the Brent price at that time is 61 US dollars/barrel. Find the optimal blending mix of these five crude oils to form 15 million barrels of this feedstock into their CDU, minimizing total cost of the inputs.

References

1. "The Economics of Process Analysis", Gregory Shahnovsky, Ariel Kigel, and Ronny McMurray of Modcon Systems, UK; Hydrocarbon Engineering, (85–92), February 2017, paper can be accessed at the link: https://www.modcon-systems.com/wp-content/uploads/2014/04/The-Economics-of-Process-Analysis.pdf
2. "Optimization Model of Crude oil Distillation Units for Optimal Blending and Operating Conditions", D. C. Lopez C., C. A. Mahecha, H. Arellano-Garcia, and G. Wozny, Ind. Eng. Chem. Research, 52, 36, (12993–13005), 2013.
3. "Mixed-integer Linear Programming Model for Refinery Short-term Scheduling of Crude Oil Unloading with Inventory Management, Heeman Lee, Jose M. Pinto, Ignacio E. Grossman, Sunwon Park, Ind. Eng. Chem. Research, 35, 5, (1630–1641), 1996.
4. "Whitepaper on Crude Blending - Honeywell Process", see https://www.honeywellprocess.com/library/marketing/whitepapers/Crude%20Blending%20Whitepaper.pdf
5. "Optimizing Chevron's Refineries", Ted Kutz, Mark Davis, Robert Creek, Nick Kenaston, Craig Stenstrom, Margery Connor, INFORMS Journal on Applied Analytics, Vol. 44, No. 1, January-February 2014, Pages 39–54, Winner of 2014 Franz Edelman Award of INFORMS.

Chapter 8
Pump Allocation at a New Zealand Oil Refinery

Shane G. Henderson, Geoffrey Gill and Parvati Patel

Abstract This chapter considers the problem of assigning a set of pumps to a set of component tanks so as to pump a prescribed volume from each component tank into a final blend tank as quickly as possible. This problem, which arose at a refinery in New Zealand, is important because delays in the blend can delay outgoing ships that take the blended product to market. Such delays incur large penalty costs. A mixed-integer programming formulation is derived that can be solved extremely quickly.

Refining NZ owns and operates the **Marsden Point Oil Refinery** in Northland, New Zealand. Crude oil is brought in by tanker, refined, and then transferred to both local and international markets. Distribution occurs by pipeline to the nearby city of Auckland, and by tanker to the rest of New Zealand and to Pacific nations.

The final step in the production process is to blend component materials from component tanks into a final product tank. The final product is then tested for compliance with certain quality requirements before being cleared for distribution. If the product is not satisfactory, then additional components (typically these are the same components from which the blend is composed), such as heavy coker gas oil and light cycle oil), are added to the product until the quality requirements are met.

It is important to ensure that the final blending step proceeds quickly. If the product is not ready when a tanker is scheduled to leave, the tanker must wait, and the refining company incurs a large penalty payment known as **demurrage**. The problem is amplified by the fact that tankers can typically only leave on a **high tide**.

S. G. Henderson (✉)
School of Operations Research and Information Engineering,
Cornell University, Rhodes Hall, Ithaca, NY 14853, USA
e-mail: sgh9@cornell.edu

G. Gill
Viva Energy Refining Pty Ltd, Refinery Rd, Corio, VIC 3214, Australia

P. Patel
Hamilton City Council, Hamilton, New Zealand

K. G. Murty (ed.), *Models for Optimum Decision Making*, International Series in Operations
Research & Management Science 286, https://doi.org/10.1007/978-3-030-40212-9_8

Fig. 8.1 Eight pumps and a blend header (that thoroughly mixes incoming components before the mixture proceeds to a final product tank) at the Magellan East Houston Oil Terminal. Image taken from https://www.global.weir/newsroom/news-articles/floway-pumps-assist-magellan/

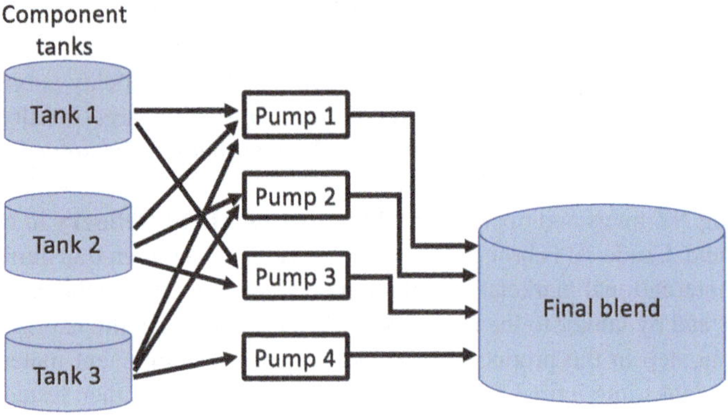

Fig. 8.2 An example with 3 component tanks and 4 pumps

Given that it is approximately 12 h between high tides, the departure delays incurred by even a small blending delay can be significant, leading to high demurrage.

The final blend is performed by pumping the appropriate quantities, as determined by solving a product blending problem (see Chap. 6), from multiple component tanks into a single final product tank. Several pumps are available to perform this blend. The photo in Fig. 8.1 shows 8 pumps and a blend header (that thoroughly mixes incoming components before the mixture proceeds to a final product tank) at the Magellan East Houston Oil Terminal.

In the contrived example in Fig. 8.2, there are 3 component tanks and 4 pumps. (There is always only one final product tank.) Each pump can be linked to different component tanks by manually opening and closing valves within the refinery. In the example, Pump 1 can be linked to Tank 1, 2 or 3, Pump 2 can only be linked to Tank 2 or Tank 3, Pump 3 can only be linked to Tank 1 or Tank 2, and Pump 4 can only be linked to Tank 3. The linking of a pump to a component tank takes nontrivial time to complete, so each pump is linked to a single component tank at the start of the blend and is not linked to a new component tank when it completes pumping the required volume during the blend. In other words, while it might be possible in theory for a single pump to be used on multiple component tanks during a single blend so as to reduce the pumping time, this is not done in practice. We therefore assume that each pump can be assigned to a tank at most once for each blend. Pumps vary according to pumping rate (measured in cubic meters per hour), and different pumps may be assigned to different tanks, including assigning multiple pumps to a single tank to combine their pumping rate so as to increase the flow rate out of the component tank. In the example, one feasible assignment of pumps to tanks is to assign Pumps 1 and 3 to Tank 1, Pump 2 to Tank 2, and Pump 4 to Tank 3. Many other assignments are possible.

The problem we discuss in this chapter is to determine pump–tank assignments that minimize the time (measured in hours) required to perform the blend. This problem is less central to the operations of oil refineries than the problems addressed in other chapters, so why discuss it? The solution to this optimization problem helps improve the efficiency of the refinery. Moreover, the solution we develop raises some important concepts in optimization that are useful in other contexts.

In the refinery we consider here there are 8 pumps and 10 component tanks, but the formulation we present can handle problems with many more pumps or tanks. Although there are more component tanks than pumps, not all tanks are used in each blend. Invariably there are at least as many pumps available as there are component tanks from which to pump, so we have a choice of how to assign pumps to tanks.

The layout of the plant and the chemical composition of the material in the component tanks preclude some pump–tank combinations. However, there is still a great deal of flexibility in making pump assignments so that the problem is a nontrivial one.

Currently, the problem is solved by individual operators who use their own (individual) ideas for determining a good pump–tank allocation. The mixed integer program (MIP) introduced here may be used to offer a suggested pump–tank allocation to operators that minimizes the pumping time over all possible pump–tank assignments.

We provide a MIP formulation for the problem of determining the pump assignment that minimizes the time required to blend the component tanks. The objective function of this MIP is nonlinear, which makes the problem harder to solve. An interesting transformation, different from those we have already seen in earlier chapters, makes the problem linear and therefore easier to solve.

An extended version of this chapter is available from the first author's website; see https://people.orie.cornell.edu/shane/pubs/pumps.pdf. Relative to this chapter, the extended version also explores the possibility of changing the volumes of components used in the blend to reduce the pumping time while still ensuring that the blend has the desired properties. Such an extension might be valuable in cases where significant demurrage costs might be incurred even when using the optimal pump assignment as derived in this chapter.

The Mixed Integer Program

Suppose that there are m tanks that contain component materials that are to be pumped into a final product tank. Tank j contains v_j cubic metres of material that must be pumped. There are $p \geq m$ pumps that can be assigned to the component tanks. (We never encountered a problem where $p < m$, i.e., where there are fewer pumps than tanks, though such a situation is theoretically possible. In such a setting at least one pump would need to be used to pump more than one component tank during the blend. Due to the rarity of such a situation, we do not consider it further here.) Let f_{ij} be the rate at which pump i can pump the contents of component tank j to the final product tank in cubic metres per hour. This rate depends on the pump/tank combinations because the pumping rate depends on both the capacity of the pump and the viscosity of the fluid contained in the tank.

Let x_{ij} be 1 if pump i is assigned to component tank j, and 0 otherwise. Pumps can be assigned to at most one component tank, so that for each pump $i = 1, 2, \ldots, p$,

$$\sum_{j=1}^{m} x_{ij} \leq 1.$$

This constraint permits multiple pumps to be assigned to a single tank.

Furthermore, certain pump/tank combinations are not possible for the reasons discussed in the introduction. This constraint may be implemented by imposing an upper bound b_{ij} on x_{ij} that is 0 if pump i cannot be assigned to tank j, and 1 otherwise.

Turning now to the objective function, let t_j denote the time required in hours to pump the requisite volume (in cubic meters) from component tank j. Given the pump–tank assignments, x_{ij}, it follows that the overall flow rate of material out of tank j in cubic meters per hour is

$$\sum_{i=1}^{p} f_{ij} x_{ij},$$

for $j = 1, 2, \ldots, m$. Therefore, the time in hours needed to pump tank j will be the required volume in cubic meters divided by the flowrate in cubic meters per hour, i.e.,

$$t_j = \frac{v_j}{\sum_{i=1}^{p} f_{ij} x_{ij}}, \tag{1}$$

for $j = 1, 2, \ldots, m$ such that the denominator is strictly positive.

The total time required for the blend is the time required for the last of the component tanks to complete pumping, i.e., the maximum of the t_j values. We seek the assignment of pumps to tanks that minimizes the maximum time required to pump any individual component tank, so we want to solve the problem

$$\min_{x} \max_{j=1}^{m} t_j.$$

This problem is nonlinear, owing to the presence of the x_{ij} variables in the denominator of (1). But an elegant transformation can be used to make the problem linear. Define $r_j = 1/t_j$ to be the reciprocal of the pumping time of tank j, so that

$$r_j = \frac{\sum_{i=1}^{p} f_{ij} x_{ij}}{v_j}.$$

The problem becomes one of maximizing the minimum value of r_j. This can be accomplished with linear integer programming through a standard approach (p. 152, [1]). In particular, if we let r denote the minimum of the r_j's, then we wish to maximize r subject to the constraints that $r \leq r_j$ for each $j = 1, \ldots, m$.

The complete formulation is as follows.

$$\begin{array}{lll}
\max r & & \\
\text{subject to} & & \\
r & \leq \frac{\sum_{i=1}^{p} f_{ij} x_{ij}}{v_j} & j = 1, \ldots, m \\
\sum_{j=1}^{m} x_{ij} & \leq 1 & i = 1, \ldots, p \\
0 \leq x_{ij} \leq b_{ij} & \text{integer} & i = 1, \ldots, p,\ j = 1, \ldots, m.
\end{array}$$

The optimal objective value r^* to this problem then yields the minimum time required to pump the tanks as $1/r^*$.

For ease of reference, the input parameters in this formulation are as follows.

- m The number of component tanks that need to be pumped to complete the blend;
- p The number of pumps that can be assigned;
- v_j The volume, in cubic meters, of material in component tank j that needs to be pumped, $j = 1, 2, \ldots, m$;
- f_{ij} The flow rate in cubic meters per hour that Pump i can pump the material in Tank j, for $i = 1, 2, \ldots, p$, $j = 1, 2, \ldots, m$;
- b_{ij} 1 if Pump i can be assigned to Tank j and 0 if not, for $i = 1, 2, \ldots, p$, $j = 1, 2, \ldots, m$.

The decision variables are

- r The reciprocal of the pumping time, in hours^{-1}, and
- x_{ij} 1 if Pump i is assigned to Tank j, and 0 otherwise, for $i = 1, 2, \ldots, p$, $j = 1, 2, \ldots, m$.

This problem can be solved using any of a variety of software tools. We used Google's OR-Tools package in Python. For a typical problem with 8 pumps and 5 tanks, the solution is obtained in milliseconds.

We now give exercises for students to solve. For each exercise, we provide the solution so students can verify whether the answer they obtained is correct.

Exercise 1 Five component tanks contain different materials, and the required pumping volumes for the 5 tanks are 2279, 590, 332, 290, and 3559 m^3. Eight pumps are available to pump these tanks, and the pumping rates and bounds are given in

Table 8.1 Pumping rates in cubic metres per hour, and bounds (in parentheses) for various pump–tank combinations. If a bound equals 1, 0 it means "*can* be assigned to this tank", "*can't* be assigned to this tank", respectively. The pumping rates vary because of the various ratings of the pumps, and the viscosity of the components held in the tanks

Tanks	Pumps							
	1	2	3	4	5	6	7	8
1	140 (1)	300 (0)	320 (0)	200 (1)	120 (1)	340 (0)	150 (0)	500 (0)
2	112 (0)	240 (0)	256 (1)	160 (1)	96 (1)	272 (0)	120 (0)	400 (0)
3	112 (1)	240 (1)	256 (0)	160 (1)	96 (0)	272 (1)	120 (0)	400 (0)
4	84 (0)	180 (1)	192 (1)	120 (1)	72 (0)	204 (1)	90 (0)	300 (1)
5	84 (1)	180 (0)	192 (1)	120 (0)	72 (0)	204 (0)	90 (1)	300 (0)

Table 8.1. (Recall that if Pump i cannot be assigned to Tank j, then we set the bound $b_{ij} = 0$. The bounds are given in parentheses in the table.) What is the optimal assignment of pumps to tanks, and what is the minimum time required to complete the pumping?

Solution to Exercise 1 *The full formulation for this problem is as follows.*

$$\begin{array}{ll}
\max r & \\
r & \leq 0.061x_{11} + 0.088x_{41} + 0.053x_{51} \\
r & \leq 0.434x_{32} + 0.271x_{42} + 0.163x_{52} \\
r & \leq 0.337x_{13} + 0.723x_{23} + 0.482x_{43} + 0.819x_{63} \\
r & \leq 0.621x_{24} + 0.662x_{34} + 0.414x_{44} + 0.703x_{64} + 1.034x_{84} \\
r & \leq 0.024x_{15} + 0.054x_{35} + 0.025x_{75}
\end{array}$$

$$\begin{array}{lll}
x_{11} + x_{12} + x_{13} + x_{14} + x_{15} \leq 1 & & \\
x_{21} + x_{22} + x_{23} + x_{24} + x_{25} \leq 1 & & \\
x_{31} + x_{32} + x_{33} + x_{34} + x_{35} \leq 1 & & \\
x_{41} + x_{42} + x_{43} + x_{44} + x_{45} \leq 1 & & \\
x_{51} + x_{52} + x_{53} + x_{54} + x_{55} \leq 1 & & \\
x_{61} + x_{62} + x_{63} + x_{64} + x_{65} \leq 1 & & \\
x_{71} + x_{72} + x_{73} + x_{74} + x_{75} \leq 1 & & \\
x_{81} + x_{82} + x_{83} + x_{84} + x_{85} \leq 1 & & \\
x_{12} = x_{14} & = 0 & \\
x_{21} = x_{22} = x_{25} & = 0 & \\
x_{31} = x_{33} & = 0 & \\
x_{45} & = 0 & \\
x_{53} = x_{54} = x_{55} & = 0 & \\
x_{61} = x_{62} = x_{65} & = 0 & \\
x_{71} = x_{72} = x_{73} = x_{74} & = 0 & \\
x_{81} = x_{82} = x_{83} = x_{85} & = 0 & \\
0 \leq x_{ij} \leq 1, \ \textit{integer} & & i = 1, \ldots, 8; \ j = 1, \ldots, 5.
\end{array}$$

Table 8.2 An optimal set of pump assignments for Exercise 1

Tank	Pump(s)
1	4
2	5
3	2
4	8
5	1, 3, 7

Solving this integer program, we get that the minimum time required to pump these tanks is 11.4 h. An optimal assignment of pumps to tanks is given in Table 8.2.

It may seem odd that Pump 6 has not been assigned. It cannot be assigned to Tank 1, which takes the longest time to pump, so even if we use Pump 6, it will not reduce the overall pumping time. An operator may still wish to assign Pump 6 to another tank, because there are uncertainties in pump flow rates due to uncertainties in viscosity of component materials and pump status. (These uncertainties arise because flow rates and the like cannot be determined with perfect accuracy. For example, in reality it is possible that during the blending operation additional component material is deposited by the refinery in the component tanks, thereby changing their chemical composition and potentially their viscosity.) Our formulation does not consider such uncertainties.

Exercise 2 In Exercise 1 above, Pump 6 is not needed in an optimal assignment of pumps to tanks. Suppose instead that Pump 6 can be assigned to any of the tanks. What now is the minimum pumping time?

Solution to Exercise 2 *We change the bounds for Pump 6 to all equal 1, so that Pump 6 can now be assigned to any of the tanks 1 through 5. The resulting solution requires 7.3 h to complete the pumping. Tank 5 now takes the longest time of all tanks to pump. Table 8.3 gives the new optimal assignment of pumps to tanks.*

Table 8.3 An optimal set of pump assignments for the modified example in Exercise 1, where Pump 6 can be assigned to any tank

Tank	Pump(s)
1	1, 4
2	5
3	2
4	8
5	3, 6, 7

Exercise 3 Consider the original problem in Exercise 1. If the required volume to be pumped from Tank 5 increases sufficiently, the minimal time to pump starts to increase. At what volume in Tank 5 does the optimal pumping time first start to increase?

Solution to Exercise 3 *This happens when the time required to pump Tank 5 equals the time required to pump the bottleneck tank (the tank that takes the longest time to pump), Tank 1. The time required to pump Tank 1 is its volume* v_1 *divided by the flow rate out. Since only Pump 4 is assigned to Tank 1, the flow rate out is* $f_{21} = 200\,m^3/h$. *The resulting time to pump Tank 1 is* $2279/200 = 11.4\,h$, *as expected, since Tank 1 is the bottleneck. Tank 5 is assigned pumps 1, 3, and 7 for a total flowrate out of* $f_{15} + f_{35} + f_{75} = 366\,m^3/h$. *It will take 11.4 h to pump when its volume* $v_5/366 = 11.4$, *i.e., when* $v_5 = 4172\,m^3$. *Rerunning the optimization problem with* $v_5 = 4172\,m^3$ *yields an optimal pumping time of 11.4 h, while rerunning with* $v_5 = 4180\,m^3$ *(for example) yields an optimal pumping time of 11.42 h. So the desired solution is* $v_5 = 4172\,m^3$.

Exercise 4 Develop a *greedy heuristic* for this problem, i.e., a simpler approach than the optimization formulation above for assigning pumps to tanks that one would expect to do quite well in general, though there are no guarantees that it would give the assignment that minimizes the pumping time. Your method should be *greedy* in that it assigns one pump at a time, without regard to the later assignments of pumps to tanks. Run your greedy algorithm on the problems in Exercises 1 and 2 and compare your result to that of the integer-programming algorithm above.

Solution to Exercise 4 *There are many possible heuristics. Here is one that, at each stage, assigns a pump in an attempt to maximally reduce the maximal pumping time. Recall that i indexes the available pumps, j indexes the component tanks with material to be pumped,* v_j *gives the volume of material to be pumped from component tank j, and* f_{ij} *gives the rate in cubic meters/hour that pump i can pump material from component tank j, for* $i = 1, 2, \ldots, p$ *and* $j = 1, 2, \ldots, m$. *Let S be the set of all feasible pump-component tank combinations* (i, j) *such that pump i can be linked to component tank j, i.e.,* $S = \{(i, j) : f_{ij} > 0\}$.

Let R be the set of remaining *pumps that are yet to be assigned and let A be the set of* assigned *pump–tank combinations. For pump j, let* t_j *be the time required to pump tank j given the pump assignments so far,* $j = 1, 2, \ldots, m$. *The greedy algorithm begins with* $R = \{1, 2, \ldots, p\}$, $A = \emptyset$, *and* $t_j = \infty$ *for all* $j = 1, 2, \ldots, m$. *It then finds the pump-component tank combination* (q, r) *that corresponds to the largest value for the ratio (cubic meters of material required to be pumped from component tank divided by the flow rate)* v_j/f_{ij} *among all* $(i, j) \in S$, *then links pump q to component tank r. Since pump q has now been assigned, it removes q from R for all t, adds* (q, r) *to A, and updates* $t_r = v_r/f_{qr}$.

Subsequently, whenever one or more pumps are unassigned, it first identifies the set of tanks T that are tied as the ones taking longest to pump given the pump assignments so far, i.e, $T = \{r : t_r = \max\{t_j : j = 1, 2, \ldots, m\}\}$. *It then identifies, from the pumps that remain unassigned, the pump–tank combination that, together with earlier pump assignments, yields the largest pumping time of one of the tanks in T. Mathematically, we choose* $q \in R$ *and* $r \in T$ *that maximizes the ratio*

$$\frac{v_r}{f_{qr} + \sum_{(s,t)\in A: t=r} f_{st}}.$$

We now remove q from the set of remaining pumps R, add (q, r) *to the set of assigned pump–tank combinations A, and update*

$$t_r = \frac{v_r}{\sum_{i:(i,r)\in A} f_{ir}}.$$

This process continues until all the pumps are assigned $(R = \emptyset)$ *or there are no remaining pumps that can reduce the pumping time of at least one tank in the set T.*

Applying our greedy algorithm to the problem in Exercise 1, we get the following table, where each row contains an iteration of the greedy algorithm. The first column gives the set T, the second gives the pump that is assigned, and the third column gives the tank to which the pump is assigned.

T	Pump	Tank
$\{1, 2, 3, 4, 5\}$	1	5
$\{1, 2, 3, 4\}$	5	1
$\{2, 3, 4\}$	4	2
$\{2, 3\}$	2	4
$\{3\}$	6	3
$\{5\}$	7	5
$\{5\}$	3	5

Thus Tank 1 is assigned Pump 5, Tank 2 is assigned Pump 4, Tank 3 is assigned Pump 6, Tank 4 is assigned Pump 2, and Tank 5 is assigned Pumps 1, 3, and 7. Pump 8 is not assigned in the last step, because it cannot reduce the time required to pump Tank 5, which is the tank that is slowest to pump. The overall pumping time is 19 h. This does not compare very well with the optimal solution that takes 11.4 h.

Now suppose that Pump 6 can pump any *tank. The heuristic then makes the same sequence of assignments, so the pumping time remains the same, being 19 h, which compares very poorly to the optimal time.*

Acknowledgements This work was conducted while the first and third authors were, respectively, a lecturer and a final-year undergraduate in the Department of Engineering Science at the University of Auckland in New Zealand. At that time, the second author was an analyst at Refining NZ. This work was partially supported by New Zealand Public Good Science Fund grant UOA 803. We thank Katta Murty for helpful suggestions.

Reference

1. Dantzig, G. B., and M. N. Thapa. 1997. *Linear Programming 1: Introduction*. Springer.

Index

K. G. Murty (ed.), *Models for Optimum Decision Making*, International Series in Operations Research & Management Science 286, https://doi.org/10.1007/978-3-030-40212-9

The manufacturer's authorised representative in the EU is Springer Nature Customer Service Centre GmbH, Europaplatz 3, 69115 Heidelberg, Germany. If you have any concerns regarding our products, please contact ProductSafety@springernature.com

Printed and bound by CPI Group (UK) Ltd, Croydon, CR0 4YY
30/09/2026
02225465-0003